RIVERS IN ROCK

Rivers in Rock: a view of Elora Gorge looking downstream from just below the Irvine-Grand confluence. Bedrock forms the channel floor, banks, and walls.

RIVERS IN ROCK

ELORA GORGE
Field Companion and Natural History

KENNETH HEWITT

WILFRID LAURIER
UNIVERSITY PRESS

Wilfrid Laurier University Press acknowledges the support of the Canada Council for the Arts for our publishing program. We acknowledge the financial support of the Government of Canada through the Canada Book Fund for our publishing activities. Funding provided by the Government of Ontario and the Ontario Arts Council. This work was supported by the Research Support Fund.

Library and Archives Canada Cataloguing in Publication

Title: Rivers in rock : Elora Gorge field companion and natural history / Kenneth Hewitt.
Names: Hewitt, Kenneth, author.
Description: Includes bibliographical references and index.
Identifiers: Canadiana 20230180655 | ISBN 9781771125802 (softcover)
Subjects: LCSH: Fluvial geomorphology—Ontario—Elora Gorge—Guidebooks. | LCSH: Geology—Ontario—
   Elora Gorge—Guidebooks. | LCSH: Natural history—Ontario—Elora Gorge—Guidebooks. | LCSH: Elora
   Gorge (Ont.)—Guidebooks. | LCGFT: Guidebooks.
Classification: LCC GB568.15 .H495 2023 | DDC 551.44/20971342—dc23Classification: LCC GB568.15 .H495
2023 | DDC 551.44/20971342—dc23

Cover design by John van der Woude, JvdW Design. Interior design by Daiva Villa, Chris Rowat Design.
The front cover image shows the waters of the Grand and Irvine Creek, which join just downstream of Elora in what is aptly known as "The Cove." The streams are overlooked by hanging rock and by solution and crevice caves, carved in massive limestone typical of Elora Gorge. Angular rockslide boulders litter the channels. Trees of the canopy, above, have roots in the same bedrock. Photograph by Kenneth Hewitt.

This book is printed on FSC® certified paper. It contains recycled materials and other controlled sources, is processed chlorine-free, and is manufactured using biogas energy.
Printed in Canada

Wilfrid Laurier University Press is located on the Haldimand Tract, part of the traditional territories of the Haudenosaunee, Anishnaabe, and Neutral Peoples. This land is part of the Dish with One Spoon Treaty between the Haudenosaunee and Anishnaabe Peoples and symbolizes the agreement to share, to protect our resources, and not to engage in conflict. We are grateful to the Indigenous Peoples who continue to care for and remain interconnected with this land. Through the work we publish in partnership with our authors, we seek to honour our local and larger community relationships, and to engage with the diversity of collective knowledge integral to responsible scholarly and cultural exchange.

The author acknowledges that Elora Gorge lies in the traditional territory of the Attawandaron, Anishnaabe, and Haudenosaunee peoples, and in the territory governed by the Dish with One Spoon wampum, the Two Row wampum and the Treaty of Niagara (1764). It is also part of the Haldimand Tract, the land promised to the Six Nations that included ten kilometres on each side of the Grand River from its mouth to its source.

Confined in a channel worn out of the solid rock to a depth in different places of from sixty to eighty feet this beautiful river has a never failing attraction, and to those who love to study nature the rocks have a wonderful story to tell.

—John Robert Connon (1862–1931)

Contents

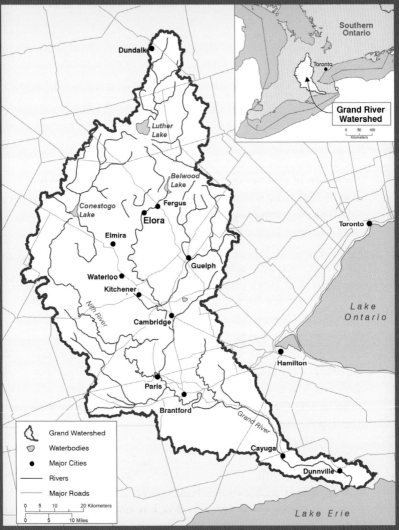

Figure 0.1   **The Grand River Basin and its setting within Southwestern Ontario.**

Preface

The gorges of the upper Grand River are tucked away in the hinterland of south-central Ontario some seventy kilometres west of Toronto, and a similar distance northwest of Niagara Falls. The main gorge begins at "The Falls" in Elora on the upper Grand River. A little way downstream another gorge, on Irvine Creek, joins the Grand. This tributary gorge has its own unique features and appeal. It stretches upward to another falls, partly engineered, at Woolwich Street Bridge in Salem. Below its junction with the Irvine, the Grand continues through the Elora Gorge Conservation Area, managed by the Grand River Conservation Authority (GRCA).

The two main canyons, and some lesser ravines descending to them, are usually treated together as "Elora Gorge." For those who live in the area they are simply "the gorge." Accompanying maps and descriptions show how to reach them and gain access (figs. 0.1 and 0.2).

Elora Gorge can seem strangely out of place. You travel to it through subdued or gently rolling terrain. The roads run straight for miles. The streams you cross meander slowly through gentle meadows and fringing reeds. Here and there they pass under low bridges. Bedrock is rarely seen. It lies buried beneath soils and glacial sediments of the last Ice Age. Surrounding Elora is an ever-growing web of residential subdivisions, shopping malls, and industrial plants. Where they have not yet succumbed to urban sprawl and industrialized

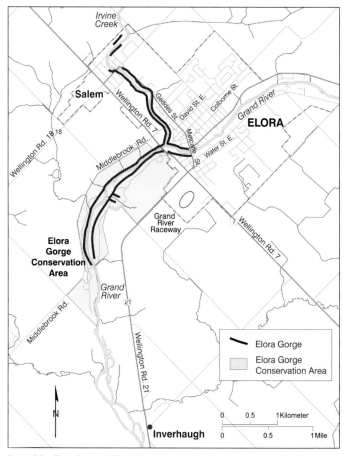

Figure 0.2   **Elora Gorge and immediate surroundings.**

paving, they back onto a rectangular patchwork of fields and woodlots, the quiet gardens of small towns.

None of this prepares you for the gorge. In the final steps to its edge the familiar landscapes of Southern Ontario are thrust aside. The land drops away sharply at your feet. Below and opposite are sheer rock walls. Down in the gorge the ruggedness seems remote from towns and cities; more than actual distances suggest. You hear the rush of water, glimpsed through the trees as it hurries among tumbled boulders or laps the mouth of a cave.

Many features would seem more at home in mountain valleys of the Canadian Rockies or other alpine wilderness, not in lowland Ontario. There is good rock here, hard and dry where exposed above the river. There are sheer cliffs for the eye to wander over. Get used to placing your hand on bedrock exposures, to note the different shapes and textures: these help identify fossil remains and reflect how the rocks formed or have been revealed by erosion.

Down in the gorge there is stillness too; quiet and mysterious places are encountered. Pause and listen, begin to catch the changing cadences of white water, the smooth sections passing into deep pools or speeding up at rocky narrows, the many springs that emerge on or below the cliffs. As familiarity grows you discover unexpected forms hidden away: a cluster of fossil shells in the rock, a delicate wildflower, animal trails and middens under **rock shelters**. You begin to sense the legacies of a rich and remarkable story; at least, I hope you will find it so.

> Terms in **blue bold** when they first appear are defined in the Glossary on page 201. Other terms of special interest, mostly scientific and technical, are in *italics*.

The distinguishing features of Elora Gorge, and the focus of this exploration of them, involve *rivers in rock*: streams whose channels, banks, and confining walls are carved in bedrock. This phenomenon defines not only characteristic landforms of the gorge, but how they respond to, and resist, erosional forces.

The natural landscape is the main interest of this book, which updates and expands an earlier work published in 1995.[1]

Before that, the last popular study was David Boyle's *On the Local Geology of Elora*, a talk he presented in 1874 when he taught at Elora Public School.[2] More recently, the work of Earth scientists in Southern Ontario has revealed much that was unknown in Boyle's time and, indeed, in 1995; essential background for what is presented here.

It must be admitted that rivers in rock have received much less study than alluvial streams, in Southern Ontario and most river basins elsewhere, partly due to difficult terrain, partly to greater commercial interests. Very little of the relevant science has been applied to the gorge. Much of the best work is only available in technical reports. I draw upon these where they shed light on the features and story of the gorge. I also present relevant ideas about rivers in rock, with some of the scientific technicalities from the Earth sciences. In particular, the study relates to landforms and the field of **geomorphology**. The aim, however, is to keep the language and explanations accessible, while trying not to ignore the complexities and many unanswered questions. There is still a need for landscape detective work, looking toward the next leaps in understanding.

The natural landscape of the gorge is my main interest, but nowhere are marks of humans entirely absent. True, during the many millennia when First Nations people came and went, they left few human-made landscape changes. In the gorge itself vestiges are few compared to tracts of the Grand to the south of Elora, where thousands of Indigenous archaeological sites are known.

A quite different story played out after the arrival of Europeans. Even in the gorge you become aware of the effects of land clearance, dams, and other control works that regulate stream flow. Amid the trees in the gorge lie abandoned lime kilns, small factories, and crumbling farm buildings.

Since the 1870s, many visitors have explored the gorge in summer. Some have come for canoeing and tubing, others to scale the rock walls. In winter there are cross-country ski trails, and a few climbing enthusiasts scale the giant icicles that drape the gorge walls. The landscape raises both natural and cultural issues, questions of stewardship. Since the Grand was designated as a heritage river in 1994 the call for sensitive treatment has been, or should be, stronger.[3]

SAFETY MATTERS AND ACCESSIBILITY

If you visit the suggested sites or follow the excursions, please take precautions. Some have been chosen because they are accessible to persons with limited mobility, but many are too dangerous for direct access without specialized training and equipment. Paths into and along the gorge require good footwear and gear suitable to weather and season. Safety concerns need special attention in winter and spring. The cliffs present some perennial dangers. Unless you are an experienced climber or accompanied by one, they are best avoided.

Some of the sites and features discussed in this book have portions that are accessible year-round. A few, mainly along the streets of Elora, are wheelchair accessible. Some have safety barriers, but particular care should be taken near cliff tops and across steep slopes. Fatal falls of people and pets into the gorge are not unknown. If you are coming from a distance, check conditions with the GRCA. Elora Gorge status updates are posted on their website.[4] Portions they manage are closed for part of the year.[5]

K.H.
Elora
May 2023

Figure 1.1
Where Irvine Creek joins the Grand, typical features of Elora Gorge dominate the scene: overhanging rock masses, a dense tree canopy, and angular boulders in the stream.

Chapter One

Introduction: Riverscapes in Rock

The streams in Elora Gorge are carved almost wholly in bedrock. They include the steeper parts of the Grand River. There are narrow canyons, hemmed in on both banks by vertical rock walls. More open reaches of the gorge tend to have one cliff set back from the present-day water's edge, but within reach of the highest floods.

———————

KEY FEATURES OF THE GORGE

The two main active canyons, on the Grand and the Irvine, create the most significant exposures of the local bedrock. Upstream are some lesser rock bed channels, above Elora on the Grand, and above Salem on the Irvine. Between Elora and Fergus, "The Quarry" is a well-known swimming hole. Here one finds easily inspected rock exposures and reminders of past industrial uses of the rock. Above Salem are limestone bluffs and low cataracts as far up as the bridge at "Portage." If less dramatic, these more subdued parts add to the varieties of landform in bedrock. They have some good spots for fossil hunting and may be found awash in floods. The combined linear extent of the streams in bedrock around Elora adds up to some twelve kilometres.[1]

It bears repeating how different the gorge rivers are from surrounding areas. In other directions most valleys are in

alluvium. The river courses are marked by extensive flood plains and river flats, by meandering streams and vegetated terraces. Bedrock is rarely exposed. The rough terrain here is made up of unconsolidated, superficial deposits, mainly legacies of the Ice Age. In these places, river work is largely engaged in mobilizing, transporting, and depositing fine-grained material, predominantly silt and sand. By contrast, the prominent landforms in Elora Gorge are cut in stone. Soft sediments are largely absent, being swept through by the strong currents. Parts of the gorge floor do have deposits of boulders, mostly limestone blocks fallen from the cliffs and moved only intermittently in floods. Such conditions help single out the story of the gorge.

This book does show how, behind often bewildering complexity and detail, the regional landscape records a large-scale unity of development. The keys lie in answers to certain major questions: What gives the gorge its distinctive appearance? Why do rock gorges exist here? What processes created—and continue to shape—the gorge? In short, the answers turn out to highlight the local bedrock, long-term tectonic developments in a billion years of crustal evolution and, thirdly, certain legacies of the last Ice Age.

The gorge itself turns out to have originated late in the history of the landscape, and relatively recently: around 10,000 years ago, near the end of the Quaternary Ice Age. It involved conditions and terrain affected by gigantic ice lobes. Meltwater floods were a major factor too, as the ice lobes waned. Meanwhile, present-day processes link the fate of rivers in rock to the stability of limestone cliffs, chemical weathering processes, and certain threats from modern human activities.

BEDROCK CONTROL

The gorge channels, and walls they undercut, draw attention to the stone itself. And the gorge turns out to be carved entirely in both a single rock type and one geological unit. Respectively, they are a type of limestone known as **dolostone** or **dolomite** rock, and the geological unit is the *Guelph Formation*, which formed more than 400 million years ago in the *Silurian* epoch.

The stone is of a sort from which humans have extracted lime for use in cement, mortar, and concrete. The remains of former lime kilns are scattered along this part of the Grand River. They, and quarries that fed them, will be encountered in the gorge, including where the waste products were dumped into the river. The local limestone has also been used extensively for buildings.

As is typical of limestones, the Guelph dolostone is quite soluble in natural waters, and to varying degrees in rain, snowmelt, ground, and soil moisture. Solution weathering creates a range of developments in the landscape, including many small caves and cavities that open up into the gorge. In all, erosion here reflects a combined effort of streamflow, **rockfall**, and dissolution of the limestone (fig. 1.2).

Learn more about dolomite and dolostone in chapter 3, pp. 63–79.

Figure 1.2   **Abundant solution and scouring forms in limestone along the floor of the gorge.**

"HANGING ROCK" AND ROCKSLIDES

Figure 1.3   **Hanging rock, caves, and springs (between *Sites 4* and *5*).**

Important as the river's work is, near-vertical cliffs dominate the scene. In sunlight, the rock walls rise cream and glowing from the gorge floor. Early and late in the day they cast deep shadows over the river, made darker by the cedar trees that frame and cling to the dolostone cliffs.

A singular observation is the extent of **hanging rock**; that is, cliffs steeper than vertical. They jut outward, leaning over the river and slopes below. Some are set well back from the existing streams (fig. 1.3). They record older, abandoned river levels. Others loom precariously over the present active gorge. Some cap incisions or notches where the river undercuts the rock walls. The overhangs involve a great variety of sculptural forms, both rounded and angular. They draw attention to peculiarities of the bedrock.

Such cliffs express the physical conditions that hold the landscape in their grip. The walls exist because of the rock's strength, its capacity to resist the stresses that might bring about its downfall. In particular, the great weight of unsupported rock in large overhangs testifies to the strength of these dolostone outcrops. Conversely, the cliffs have arisen through the power of erosion in the gorge. Rockfalls and **rockslides** record past onslaughts by the river, and how the stone has eventually been pulled down by gravity (fig. 1.4).

Figure 1.4   **Rockfall on the Grand, below overhanging gorge walls, seen from "High Lookout"** *(Site 11).*

Find examples of hanging rock on Excursion 1, *Sites 2, 4, 5*, and *7*; Excursion 2, *Sites 9* and *11*; and Excursion 3, *Sites 22–23* and *29*. Find examples of rockfalls and rockslides on Excursion 1, *Sites 2–5* and *7*.

THE GRAND RIVER

Streamflow in the gorge and its power to erode presently depend upon the size and attributes of the upper Grand Basin, and the regional climate (fig. 1.5).[2] The Grand is the largest river basin in Southwestern Ontario. It drains some 6,750 km^2 of terrain to the west of the Niagara Escarpment. The highest, northern part of the basin, lying above the gorge, comprises roughly one-sixth of the whole catchment. It is the most humid part, with annual precipitation averaging over 1,000 mm. The share that falls as snow is highest in the northern and western parts of the basin. Snowfall amounts are sensitive to the extent of frozen area of the Great Lakes in winter, especially of Lake Huron. Higher amounts of evaporation occur over ice-free water and give rise to heavier "lake effect" snowfall.

Year-to-year variations are quite marked. In recent decades the amounts and shares of snowfall and rain have been changing. The incidence of thunderstorms and tornadic storms has been increasing.[3] Snow amounts, and duration of snow on the ground, were relatively high through the 1970s. Intense and months-long freezing were common in the gorge, bringing great buildups of icicles along the walls. There were efforts to promote cross-country skiing in the conservation area. However, through the 1980s and '90s snowfall declined and became less reliable. "Green spells" in the heart of winter became common, bringing episodes of melting and temporary loss of the river ice and icicles. Cross-country skiers were discouraged. Climate change was making itself felt down in the gorge.

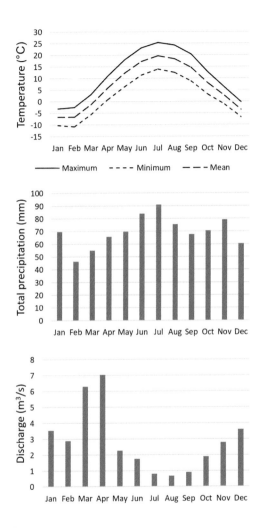

Figure 1.5   **Monthly maximum, minimum, and mean temperatures, and average monthly precipitation at the Elora weather station, 1986–2006; and average monthly stream flow for Irvine Creek at Salem, 2006–2020 (Data sources: Environment & Climate Change Canada [temperature and precipitation]; Water Survey of Canada [stream flow]).**

Some human interventions have had direct and noticeable impacts on the flow of the Grand into the gorge. These interventions include intensified land use and surface drainage, tending to increase and concentrate runoff into the streams. The Shand and Luther dams hold back the flow of the Grand from upstream. They have significant effects through flood control. In summer, low flow augmentation is important. Artificially stored water is released to maintain flows, especially for municipalities in the lower basin.

SEASONS OF THE GORGE

Much of what you encounter in the gorge depends on the weather and the season. Most visitors come in summer, preferring clear and warmer days. At these times there are usually low flows along the gorge. Other seasons may attract a few ice climbers or cross-country skiers, some anglers, birdwatchers, or butterfly collectors. For the most part the gorge is left to its own devices. However, some of the most important conditions that shape the landscape occur in the neglected seasons: during snowmelt and **ice jam** floods in spring, or tornadic and even hurricane-generated storms in summer. The streams and cliffs, of course, are exposed to the weather and runoff 365 days of the year. It is necessary to be aware of changing processes that affect the landscape alongside the growth and fall of leaves; the frost and snowdrifts of winter; the mists of fall; and the rains of spring and summer thunderstorms.[4]

For much of the winter the gorge walls are draped with huge icicles, beneath which springs continue to flow, freeze, and add to the ice (fig. 1.6). The stream bed is snow-bound or buried under slabs of ice. The snow can be soft and deep, the cold seemingly more intense in the gorge.

Perhaps the deepest impressions come with the grey mists of spring and fall, which gather to conceal the scene down in the gorge, and to moisten the walls before freeze-thaw cycles

Figure 1.6   **Winter along the Grand Gorge.**

Figure 1.7 **Fall mists and fog along Irvine Creek.**

come to attack them (fig. 1.7). Nothing quite compares with fall colours, as the maples, birches, and other deciduous trees add a bright trim to the limestone bluffs (fig. 1.8). Fallen leaves sprinkle the dark cedars and gorge floor. Attention to the tree cover introduces another surprise.

Figure 1.8   **Fall foliage along Irvine Creek gorge upstream of** *Site 5.*

THE GIFT OF TREES

At present Elora Gorge is enveloped by trees. They cloak the walls and rise well above the rim of the canyons. It may, therefore, be a surprise to find how recent is the present situation.

Figure 1.9   **Trees, mainly cedars, shroud the gorge walls along Irvine Creek.**

The scene might well appear to be "natural," an original wilderness with ancient growth intact. It is not.

By the late nineteenth century, following extensive logging of forests, land clearance, and settlement, most of the tree cover was lost in and around the gorge. We are fortunate to have a rich photographic record, thanks mainly to Elora residents and pioneering photographers, Thomas and John Connon, a father and son. Most of the landscape photographs illustrating John Robert Connon's book, *Early History of Elora and Vicinity* ([1930] 1975) show treeless slopes, often bare rock to the gorge rim.[5] Many features discussed here, from "The Cascade" to "The Tooth of Time," were severely denuded of vegetation (see fig. 6.15 in chapter 6, page 127). The gorge appeared more like the desert walls of the Grand Canyon than the boreal forest (fig. 1.10).

Figure 1.10   **The Grand River at Elora, 1892, by John R. Connon. Courtesy of Archival & Special Collections, University of Guelph, Connon Collection, XR1 MS A114300.**

Figure 1.11   Tree roots covering and growing into exposed fissures in bedrock along the gorge.

In the first half of the twentieth century, however, much of the lost woodland began to recover, along the gorge at least. This was partly deliberate policy, in the case of a few plantations of conifers. Volunteers began planting trees along some pathways and in Elora's Victoria Park. But mostly the trees came back as economic depression gripped the region, an outcome of the world wars and Great Depression, along with centralization of the Ontario economy. Local factories and a power plant opposite the Elora Mill were closed. Farms were abandoned. The trees were able to grow back largely unmolested. Eventually they would hide the abandoned farmhouses, orchards, and grazing lands of the early settlers. One can only marvel at how well the trees have recovered. They adorn most of the gorge and appear as a defining legacy, shading the otherwise bare rock and stony paths. Their roots are a critical factor in the stability or weakening of slopes (fig.1.11).

In effect, the gorge scenery and conservation area are a gift of the returning trees, evidently helped by the small, hardy cedars that had survived in hard-to-reach parts of the gorge walls. The eastern white cedar seems equally at home with its roots in swampy ground or clinging to fissures at the head of vertical rock faces, in limestone terrain. Studies on the nearby Niagara Escarpment have shown them to include the longest-lived trees discovered in Ontario, some over a thousand years old.[6]

OLD LAND, YOUTHFUL LANDSCAPES

Landform studies refer to time as both actual dates and chronologies. Landscapes have been popularly characterized as being in infancy, youth, middle, or old age. In the early days of landform science this was a common way to describe the terrain. Steep slopes suggest lots of activity and the erosional energy of youth; subdued ones the slower, weaker struggles of old age.

More recently, this language has been questioned as hardly scientific, but it can still offer useful teaching moments and a path into the greater complexities for newcomers to the field. Different ages serve to emphasize how the keys to landscape are always evolving and are linked to duration or stage of evolution. Surface features can be transformed in a few human lifetimes. However, as in the gorge, legacies of distant eras are exposed next to very recent ones. Important features prove to be products of ancient geological epochs, others much more recent. The outcomes can also be deceptive, in the gorge and across Ontario.

The topography of Southern Ontario is mostly subdued. The bedrock, while acting as the immediate and critical control in many parts, is ancient—hundreds of millions of years older than the gorge![7] And long before the gorges were cut, large parts of the region's bedrock had been stripped away, the land surface coming close to the ultimate subdued landforms of old age, examples of the "end game" of land surface erosion as indicated by the vast lowlands around Hudson's Bay.

On a human scale, however, it is the ruggedness of the Canadian Shield that is impressive enough. So are bedrock landforms like Niagara Escarpment and Elora Gorge. It appears the regional landscape here resembles an "old land," but refreshed in places by more recent events, temporarily rejuvenated.

This directs the inquiry toward the pivotal question: how did the gorge originate? The most likely view, presented in chapter 5, is that the gorge owes its existence to the Quaternary Ice Age. The canyons came into existence during the final phases of the last glaciation, roughly 10,000 years ago. Thus it turns out that the gorge is not only youthful in appearance but much younger than the rocks in which it is carved.

As such, it would be hard to overestimate the role of the ice in the rejuvenation of the terrain.

Before getting into these more complex interpretations, however, it is important to recognize and appreciate what is encountered on the ground; to focus on the landscapes encountered and immediate questions that they pose. First-hand experiences of the cliffs and caves, cataracts and rock-slides, are recommended.

Figure 2.1
The channel of Irvine Creek is cut entirely in bedrock from David Street Bridge *(Site 5)* to "Lover's Leap" *(Site 7)*, where it joins the Grand.

Chapter Two
First-Hand Experience

The focus of this book is the landforms of Elora Gorge, and such background as will help the reader to appreciate and interpret the features they encounter. Selected sites and excursions serve to reveal salient features and questions directly encountered in the terrain. The aim is also to develop a sense of the layout and the range of landforms encountered, and the singular developments that involve rivers in rock.

————

CLASSIC VIEWPOINTS

Three classic sites are suggested as good places for the newcomer to begin: "The Falls,"[1] "Lover's Leap," and the David Street Bridge. These form a convenient triangle, easily reached from the centre of Elora and the Centre Wellington Municipal Offices (fig. 2.2, *Sites 1, 2,* and *3*). There is limited parking in Victoria Park but you can catch glimpses of the exceptional townscape along Smith, James, and Henderson Streets.[2] All three viewpoints are accessible by wheelchair, though assistance may be needed to manoeuvre over stony ground and tree roots. The viewpoints should be avoided in winter snow and ice conditions.

The views are "classic" in having been visited and praised by many over the years. Some have inspired poems, acquired

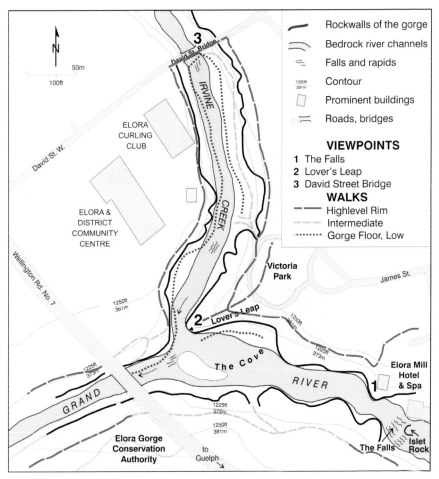

Figure 2.2   **Three "classic" viewpoints of the gorge in Elora.**

names tied to local history, and explanatory plaques. Here they serve to introduce typical features and as keys to land-form questions addressed systematically in the later chapters.

Viewpoint 1: The Falls and Islet Rock

> *"the reason for Elora's existence was the Falls on the Grand River. There was no other reason."*
> —John Robert Connon (1930)[3]

"The Falls" are a definitive feature of the Elora townscape. There are various places to observe them. The site chosen (*Site 1*) is at the northeast corner of Victoria Park, and just down-stream of the Mill Inn. The Inn overlooks the Falls and is an imposing stone structure, originally built in 1843. In former times it housed a grist mill, sawmill, and distillery, and was converted into a hotel in 1975. There have been major additions in the present century, including many windows looking out over the Falls.

Two sharply contrasted scenes are evident here, one domi-nated by human interventions, the other by natural history. Immediately downstream, the Grand enters a seeming wilder-ness, a wooded landscape of swift water and sheer rock walls. All seems to belong to a riverscape and records a story in lime-stone cliffs, rapids, rockfalls and caves. From the plunge pool at the base of the Falls, and on downstream, the gorge walls include many examples of hanging rock.

How different is the world above the Falls. Houses, shops, and abutments crowd the riverbanks. Stone and concrete walls confine the river channel. The stream descends toward the cataract through a scene that reflects some 180 years of Euro-pean settlement and industry. Bedrock is generally hidden, buried under sites of industry or commerce. This scene reflects what drew settlers to this place: convenient river crossings

Figure 2.3 **The Falls at Elora, where the Grand River divides around Islet Rock and plunges into the gorge, leaving the heavily built-up valley above** *(Site 1).*

Figure 2.4   **Islet Rock in winter, with concrete abutment on upstream side.**

and water power. As well as energy from the Falls, timber was harvested from formerly vast forests, eventually supporting a local tradition of furniture-making. As the forest was cleared and replaced by cultivation, Elora became a market and mill town. Starting in the second half of the nineteenth century, the gorge began to attract tourists.[4]

The Mill overlooks another unique feature in the river, "Islet Rock," also known locally as "The Tooth of Time" (fig. 2.3). It has served as a symbol of Elora and was for many years the official emblem of the village. Note how precariously Islet Rock seems balanced above the cataract. Conditions along the line of the river's flow show how narrow the islet's base is: a small, overhanging bedrock mass, in turn supporting a little grove of cedars and wildflowers. In floods the islet disappears in a maelstrom of white water.[5] It survives in part due to human intervention. The upstream flank is armoured by a concrete abutment, to protect against the spring melt and ice blocks that rip through (fig. 2.2).

The Falls are pivotal for the present and past evolution of the main gorge. The steep step at the Falls is where headward erosion of the main gorge presently culminates. The Falls have migrated upstream to the present location. The height, elevation, and rate of retreat of the Falls help control the development and intensity of gorge erosion. Left to itself it will continue to cut back.[6]

Viewpoint 2: Lover's Leap and the Cove

A second site overlooks the gorge from where the Grand and Irvine Rivers meet in Victoria Park (fig. 2.5; *Site 2* in fig. 2.2). From the park entrance, bear left and follow the path sloping toward where the two streams join. There you reach a narrow, walled enclosure poised vertically above the waters.

Standing at the outer wall you can take in "The Cove," a

Figure 2.5   **Lover's Leap,** viewed in snowy weather from "The Cut" *(Site 9)* across the waters of the Cove.

wide, watery space hemmed in by dolostone cliffs. Here you can find other singular and dramatic introductions to the gorge, and the theme of rivers in rock. The Cove is surrounded by caves, alcoves, limestone grottoes, and galleries, all reflected in the waters of the river. Vertical chimneys and fissures open up in the walls. The landforms derive from the combined work of undercutting by the river, dissolving of the limestone, fracturing of over-stressed rock, and the junction of two rivers.

Hanging rock leans far out over the water on all sides, and notably at the lookout point itself (fig. 1.1 in chapter 1). This feature is also called "Lover's Leap," said to commemorate an Indigenous princess who jumped to her death here after her beloved was killed in battle. The tale likely has more to do with Victorian melodrama and a desire to attract tourists than with Indigenous peoples' history.

River junctions are critical sites in landform development. The Irvine-Grand junction is of special interest.[7] Some commentators suggest that the Cove represents earlier stream captures of the Grand by Irvine Creek, or the reverse. At this particular junction there are intermittent buildups and reworking of alluvial bars and delta-like deposits. Some support trees that grow on the valley floor. The stony delta reflects the interaction of variable flows and sediment delivery of the two rivers. Contrary to expectation, deposition is more extensive along the Grand. Although it drains a larger and higher area, it is more affected by human interference, especially two dams upstream. The flow and sediment transport by the Irvine are more in keeping with natural stream flow and climatic controls. Riverbed features are changeable but tend to reflect recent weather, inter-annual conditions, or flood events.

The combined waters of the two rivers escape from the Cove under the high span of Wellington Road #7 Bridge. There, the river enters an impressive chasm that runs for several kilometres through the conservation area (see Excursion 2).

Viewpoint 3: David Street Bridge

The third viewpoint is on the rim of the Irvine Gorge, from the David Street Bridge (see fig. 2.4). The bridge can be reached by a footpath along the edge of the gorge northeast of Lover's Leap. Alternatively, you can drive to the community centre on David Street and park a few metres from the bridge. The site

Figure 2.6   View upstream along the Irvine Creek, from Victoria Park to David Street Bridge *(Site 3)*.

is wheelchair accessible from the parking lot, and the bridge
has a pedestrian sidewalk with a curb. You should still be
cautious of vehicular traffic, however. Indeed, looking down
into the deep chasm underneath the bridge can generate feel-
ings of vertigo.

Downstream of the bridge, marked differences arise
between the river's right and left flanks. The former hugs close
under vertical cliffs. The latter has an almost unbroken series
of rockslides shrouded by trees and backed by irregular cliffs.
At river level, large slabs of rock are piled up across half the valley floor. They force the water toward the right wall. This causes
erosion at the base of a wall of thin-bedded limestones. These
features are addressed in more detail in the first excursion.

The David Street Bridge of today is the latest of several
at this site.[8] One may ask, why just here? On one hand, the
bridge takes advantage of the pre-existing landforms. The
geology and landscape history made this the narrowest
place for crossing between the precipitous walls of the Irvine
Gorge. The bridge sits upon promontories of hanging rock
that have resisted the collapses and rockslides seen both up-
and downstream. On the other hand, however, the rock here
also has the same fractures as those identified with the rockslides from the adjacent walls. Expect a collapse at this bridge
too, some day!

EXCURSIONS

The excursions help to add a greater sense of detail and associa-
tions of the gorge scenery. Parts of each excursion give access
to the various levels that make up gorge landforms. Paths following the rim reveal more of the variety of rock walls and
hanging rock, how they relate to the surrounding terrain, and
vertical relations to the gorge floor. A few paths negotiate the
steep mid-levels between gorge rim and river, revealing what
occurs amid the trees and former higher channel locations of

the river. Some paths take you down to the gorge floor and provide closer views of rivers in rock, a chance to explore hands-on evidence of fossils and the varieties of dolostone. There are perspectives on the cliffs from below.

> Twenty-nine sites of special interest are described in this chapter. They are numbered sequentially and shown on the accompanying maps, starting with the Falls (*Site 1*) and ending with Eyrie Viewpoint (*Site 29*).

Excursion 1: Lower Irvine Creek

The first excursion is the shortest, but is nonetheless rich in detail. A relatively limited stretch of Irvine Creek provides a microcosm of Elora Gorge landforms between "Lover's Leap" (*Site 2*) and David Street Bridge (*Site 3*). The Irvine itself flows in and over bedrock. Vertical and overhanging walls have been the sources of multiple large rockfalls and rockslides that cover half the valley floor. The many exposures of bedrock expand the sense of diverse patterns and forms in dolostone. Along the various footpaths are well-preserved examples of most Guelph Formation fossils.

The excursion can begin at the northwestern corner of Victoria Park (*Site 4*). A flight of stone steps descends from the upper rim to mid-level features of the gorge. These steps are uneven and steep. They preclude wheelchair access, but a path along the rim from Lover's Leap to the David Street Bridge provides many views down upon the features described. Note that the gorge here is treacherous in winter and closed off except for ice climbing activities.

As you descend the steps, you are quickly surrounded by cliff faces and rock features close enough to touch. Spring-fed streams once came from tiny caves or fissures and still reappear in wet weather, demonstrating the porous nature of limestone.

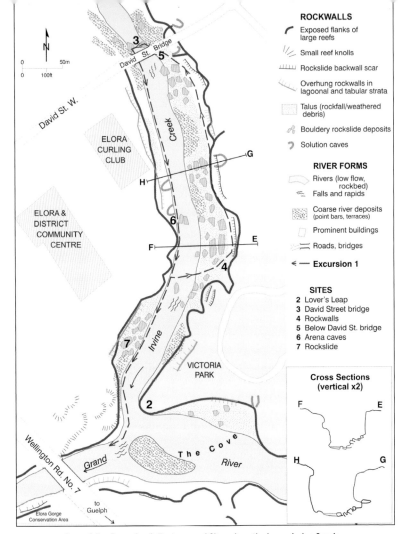

Figure 2.7   Excursion 1: Features and Sites along the Lower Irvine Creek.

The legend of the figure reads:

ROCKWALLS

Exposed flanks of large reefs

Small reef knolls

Rockslide backwall scar

Overhung rockwalls in lagoonal and tabular strata

Talus (rockfall/weathered debris)

Bouldery rockslide deposits

Solution caves

RIVER FORMS

Rivers (low flow, rockbed)

Falls and rapids

Coarse river deposits (point bars, terraces)

Prominent buildings

Roads, bridges

◄— **Excursion 1**

SITES
2 Lover's Leap
3 David Street bridge
4 Rockwalls
5 Below David St. bridge
6 Arena caves
7 Rockslide

Cross Sections (vertical x2)

Rockfall Country

At the foot of the first steep flight of steps (fig. 2.7, *Site 4*), the scene ahead is dominated by large, angular boulders. The blocks seem like giant headstones in an abandoned graveyard, monuments leaning haphazardly against each other (fig. 2.8). Many tree trunks emerge between the blocks. On cloudy days the broken rock and trees merge into the gloom. By contrast, in fine summer weather the trees and dolostone blocks offer welcome shade. But how to explain this superabundance of limestone boulders littering the flanks and floor of the gorge?

Figure 2.8   **Rockslide boulders below the Victoria Park steps in Irvine Creek** *(Site 4)*.

In fact, this is an impressive introduction to the role of slope processes in the erosional history of the gorge.[9] Evidently, countless slabs and blocks of dolostone fell, slid, or tumbled from the rock walls above to bury the lower half of the slope.

Staying on the mid-level path you pass the heads of a dozen sizable rockfalls and rockslides (fig. 2.7, cross-section E–F). You can observe how the gorge walls were undercut by the river and prepared for collapse by stress-release fractures that developed in the rock.[10] In chapter 7, I suggest that most of the rockslides along the left bank of the Irvine occurred in a single episode, triggered by a distant earthquake.[11]

It is also notable how rockfalls and rockslides bracket this whole excursion. At the upstream end (*Site 5*), immediately above the central pillar of David Street Bridge, several large blocks have fallen into the river. They tend to protect the pillar from erosion. Meanwhile, the downstream limit of the excursion (*Site 7*) is marked by a single large rockslide (figs. 2.9 and 2.10). In the 1970s it descended from the right bank below the community centre arena in what is called a *rotational slump*.

Figure 2.9
Rockslide below the community centre arena *(Site 7)*, which occurred during the spring thaw in 1973. Horizontal slabs collapsing from the wall above were released along tensile fractures in the unsupported cliff face. As they spread outward and down into the Irvine Gorge, the slabs broke up and rotated.

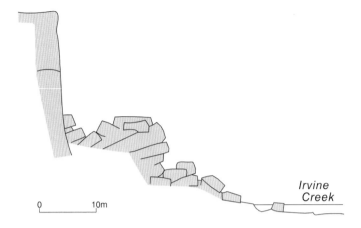

Irvine Creek

0   10m

Figure 2.10   **Cross-section of the rockslide shown in figure 2.8.**

Learn more about rockfalls and rockslides in the gorge
in chapter 7, pp. 141–44.

From the base of the first set of steps at *Site 4*, you can proceed upstream toward *Site 5* along a mid-level path that hugs
the gorge wall (see fig. 1.3 in chapter 1, page 10). Some forty
metres along this path there is a sudden, marked change. A
great promontory of massive rock overhangs the path. This is a
remnant of the older walls on either side that collapsed to generate the rockfalls. Notably, however, above the path here is a
deep, near-vertical fracture in the cliff face. This is an example
of what are called **crevice caves** on the Niagara Escarpment,
where they are common. The rock mass has already opened
up and slid down the line of fracture in the rock wall, surely
prefiguring a future collapse and massive rockslide.

Learn more about crevice caves in chapter 7, pp. 144–46.

Caves and Other Karst

Adjacent to the crevice cave are some springs that flow out at the base of the cliff from an irregular cavern. This cavern (which intersects cross-section H–G in fig. 2.7) illustrates another special role of the dolostone: it tends to dissolve in groundwater and other sources of moisture, giving rise to distinctive **karst** landforms, including deposits of *calcium carbonate*. In the mouth of this cave are typical examples. Moreover, this cave is one of more than ten encountered in this excursion, each large enough for a person to crawl inside.

Intriguing is how a line of fracture in the bedrock crosses the riverbed from this cave to others on the opposite bank at *Site 6*. The line of weakness may represent tectonic movement in the bedrock or a settling fracture in the early story of the limestone. Either way, it suggests a common link between the caves.

It is also notable that most of the larger caves enter the gorge several metres above present-day river level. This may record earlier drainage conditions with higher water tables or, perhaps, an accelerated erosional lowering of the river level since then.

Learn more about karst in chapter 8, pp. 153–69.

Massive Reef Rock

This excursion is also ideal for starting to explore the relationships between the landforms and fossils. Closer inspection of rock exposed between *Sites 4* and *5* reveals many good specimens of gorge fossils, including a variety of shellfish, corals, and the all-important sponges. In the cliffs below and downstream of David Street Bridge are horizontal banks of clam shells, ancient fossils indicated by countless small, airfoil-like shapes. In chapter 3 I show that, while dolostone is quite common in Southern Ontario, the fossils in the gorge have a special origin that is reflected in the nature of the stone itself.

These massive bedrock structures decide where the river is steeper, the locations of narrows and cataracts, the incidence of rockfalls and caves, cliff forms and spring lines. It turns out that the massive, erosion-resistant rock, notable in overhangs, was constructed by reef-building organisms on the floor of a tropical sea. It is necessary to explore further the nature of these fossils that accumulated in a **barrier reef** setting.

> Learn more about reefs and fossils in the gorge in chapter 3, pp. 71–78.

Irvine in Rock

From most parts of the mid-level path between *Sites 4* and *5* you can descend to the riverbank, where the Irvine stream flows almost entirely in bedrock. It is forced against the right bank by the rockslide boulders from the left. Agile visitors can usually get good views up and down the valley floor from boulders and stepping-stones in the stream. In summer you can usually paddle or wade across the shallows. If you reach the other side, you can make your way down the right bank of the Irvine. Between the David Street Bridge (*Site 5*) and the solution caves in the gorge walls at *Site 6*, you can walk on the smooth, rocky surface of the gorge floor. Below *Site 6* you pass the exposed flanks of some massive reefs. You follow a path over a coarse gravel terrace to *Site 7*, where you encounter the rotational rockslide mentioned earlier, which helps to push the river over to the Lover's Leap side. A little farther along, at the junction of the Irvine and the Grand, you pass under a massive reef with an astonishing depth of undercutting. Here you have a fine view of Lover's Leap (see fig. 1.1, page 6) and up the Grand across the smooth waters of the Cove to the Falls and the Mill Inn.

Back at *Site 4*, the steps and path down to the river from Victoria Park give hints of some human interactions with the gorge, including certain hazards. A small bridge once crossed the Irvine, linking Victoria Park to the community centre. The bridge was destroyed by flood waters from Hurricane Hazel in October 1954. The two concrete anchors of the bridge can be seen in the river, unmoved from where Hazel left them.

Learn more about the impacts of Hurricane Hazel and other extreme events in chapter 6, pp. 118–19, and human activities affecting the gorge in chapter 9.

Excursion 2: The Upper Grand Gorge, Elora to "Middle Bridge"

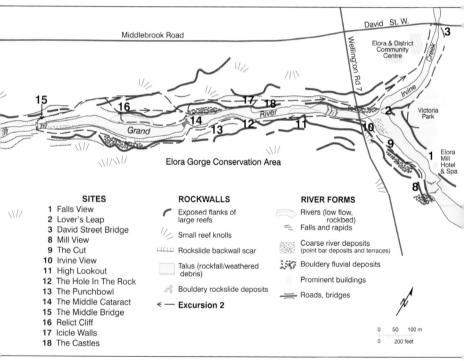

SITES
1 Falls View
2 Lover's Leap
3 David Street Bridge
8 Mill View
9 The Cut
10 Irvine View
11 High Lookout
12 The Hole In The Rock
13 The Punchbowl
14 The Middle Cataract
15 The Middle Bridge
16 Relict Cliff
17 Icicle Walls
18 The Castles

ROCKWALLS
Exposed flanks of large reefs
Small reef knolls
Rockslide backwall scar
Talus (rockfall/weathered debris)
Bouldery rockslide deposits
← — **Excursion 2**

RIVER FORMS
Rivers (low flow, rockbed)
Falls and rapids
Coarse river deposits (point bar deposits and terraces)
Bouldery fluvial deposits
Prominent buildings
Roads, bridges

0   50   100 m
0   200 feet

Figure 2.11   **Excursion 2: Features and Sites along the Grand River between the Falls and Middle Bridge.**

This somewhat longer itinerary brings greater awareness of the scale and scope of the main Grand Gorge. It takes advantage of various paths between the Falls at Elora (fig. 2.11, *Site 8*), and the Middle Bridge in the conservation area (*Site 15*). The excursion can begin opposite the Mill Inn (fig. 2.12). Visitors in the conservation area may reverse the order, beginning by following roads in the conservation area to the Middle Bridge (*Site 15*). Paths on both sides of the river can be followed. It is advisable to get the latest information on access and conditions from the GRCA website.

Figure 2.12   **View of Grand River approaching the gorge beside the Mill Inn, Islet Rock at lower left.**

If you commence at Elora, *Site 8* gives another view of the Falls from opposite the Mill Inn, in the midst of what had been an industrial site. There used to be a large furniture factory here, and a flume leading to Elora's original hydroelectric plant, closed many years ago and now replaced by a more modern version.[12] After a series of short bedrock steps the river channel plunges between massive rock towers into the main gorge. Suddenly the walls rise vertically from the stream. The exposed flanks loom over the gorge threshold like stone guardians (fig. 2.13).

Figure 2.13   **The "Stone Guardians," looking downstream through the narrows below the Falls *(Site 1).***

The left bank paths between *Sites 8* and *10* proceed through tumbled terrain opposite Lover's Leap (*Site 2*). Immediately below the Falls (*Site 1*), the gorge walls consist of a series of large, smooth, semicircular depressions. They are remains of giant whirlpools and **potholes** marking where and how the Falls cut back through this section, retreating headward. A series of rocky depressions and knolls crowns massive reef formations. These seem to have been eroded and exposed at an earlier time and by several different river channels that once passed this way.

Figure 2.14   **The Grand-Irvine junction and Wellington Road #7 Bridge, viewed looking downstream from above the Cove.**

About 100 metres downstream from the Falls is a narrow gully in the rock, "The Cut" (*Site 9*). A path slants down to the river and a triangle of gravel beach. Here you stand between overhanging promontories and deep recesses in bedrock. An umbrella-like shelter on the upstream flank is the most extreme overhang in the whole gorge. There is also a panoramic view of the junction of the Grand and Irvine (fig. 2.14).

In the conservation area, the excursion can begin on the left bank of the Grand, below the Wellington Road #7 bridge. The gorge here is a relatively narrow, nearly straight canyon

Figure 2.15   **Sheer and overhanging dolostone reef mass and rockfall, viewed from "High Lookout"** *(Site 11).*

in bedrock (*Sites 11–14*). It continues so almost as far as the Middle Cataract. The cliffs are distinctly higher than in the Cove or Lower Irvine Gorge. The river is twice as wide. Above the rim of the main gorge, many low rock terraces are encountered, and relict cliff lines. John Connon, writing in 1930, described them as "ledges of rock, in some places as wide as a street."[13] They represent old stream levels, clues to the early evolution of the gorge.[14]

On the left bank, about 200 metres downstream from the Wellington Road #7 bridge, is "High Lookout" (*Site 11*). It is an ideal spot to take stock of the scale and features of one of the most majestic parts of Elora Gorge. On the walls opposite, cream-coloured hanging rock marks the sites of large ancient reefs (fig. 2.15). On the far side of the gorge are also many springs that emerge relatively high up at *Site 17*. The springs mark where a higher water table surrounds the inflow of waters from a **hanging tributary**. In winter, a few days of heavy frost are enough to cover the whole wall with magnificent displays of icicles, reaching from top to gorge floor (fig. 2.16).

Figure 2.16   **"Icicle Walls" near *Site 17*.**

Learn more about reef masses in chapter 3, pp. 66–71; hanging or discordant tributaries in chapter 6, pp. 126–29; and the significance of "Icicle Walls" in chapter 6, pp. 122–25.

A further 200 metres along the path is the "Hole in the Rock" (*Site 12* and Fig. 2.17). A flight of steps disappears suddenly into the earth. Once you get used to the darkness under the shadow-filled roof and walls, or aided by a flashlight, well-preserved fossils are revealed. There are clusters of lamp shells (*Brachiopods*) and sponges interspersed with coral colonies. The latter include elaborately patterned *Tabulates*. The "Hole" emerges on the other, riverward flank under a natural arch and a low stone terrace. To the left is the famous "Wampum Cave," so named because Indigenous relics were found here in the 1880s by David Boyle's students from the Elora School. Below the arch is a rock-fall. It represents a large collapse of strata that once formed a natural overhang. Now it supports sprawling cedar tree roots.

Figure 2.17   **"Wampum Cave" and the "Hole in the Rock," by John R. Connon. Courtesy of Archival & Special Collections, University of Guelph, Connon Collection, XR1 MS A114305.**

Learn more about fossil shellfish and corals in chapter 3, pp. 71–77, and fossil hunters in chapter 9, pp. 173–77.

Some fifty metres further along the same path you find remnants of a former river channel, now dry. It entered the gorge about twenty metres above the present stream level (*Site 13*). The river used to descend a short waterfall into a whirlpool. Here it carved out "The Punchbowl," a wide, circular depression. This is another reminder of the complicated history of former rivers criss-crossing this area, now long lost. It is another good site to look for fossils. At least one coral colony here is half a metre in diameter.

Near the Punchbowl is a first view over the Middle Cataract, also known as "The Little Falls" (*Site 14*). The channel steepens into a hundred-metre stretch of cascading water. The river rushes through a series of reefs exposed in the limestone and on into narrow plunge pools (fig. 2.18). This is a favourite spot for "tubers" (recreationists riding inflated inner tubes) to get into the water. On the far bank a metal stairway aids access to the water. An extensive series of irregular **strath terraces** lies above the vertical and overhanging walls—a multi-storied record of stream incision in bedrock.

Already flowing rapidly, the Grand steepens even more, and some 250 metres farther downstream the river enters another narrows and an impressive cataract, where the Middle Bridge spans a narrow slot (*Site 15*). Access to *Site 15* from the right bank allows you to inspect an intriguing variety of potholes and other scoured bed forms. From here, Excursion 2 returns upstream along the right bank of the Grand from Middle Bridge back to the Irvine Junction. Note that the loop of Excursion 3 also begins and ends at Middle Bridge.

Learn more about potholes and other scoured stream bed forms in chapter 6, pp. 110–14.

Heading upstream from the Middle Bridge, the park road takes you to a good viewpoint looking down on the Middle Cataract. You have several choices of route from here: stay

Figure 2.18 **The Middle Cataract or "Little Falls"** *(Site 14),* where the floor of the Grand plunges through outcropping reef limestone, creating multiple potholes and other erosion features in bedrock.

at this mid-level, return to the road above, or go down to the river. The latter is made easier by a flight of metal steps that take you to the gorge floor at *Site 14*. In the low cliffs overlooking the cataract, many fossils are exposed. They include colonies of a characteristic coral of the Guelph Formation, *Fletcheria*—called *Picnostylus* in older texts (see fig. 3.5, page 72, for illustrations).

Around 250 metres along the conservation area road, heading upstream from the Middle Bridge, a narrow path on the right descends through the trees, down to and under a low, *relict cliff* with overhanging slabs (*Site 16*). Marking an old level of the gorge, the cliff preserves its steepness and provides a sheltered spot that can be delightfully warm and sunny through morning and midday, even in winter. There appears to have been an *inter-reef area* here, with beds of variable thickness full of **vugs** (small cavities), and a ramp of weathered debris a little out from the cliff, sloping down to a swampy terrace at the old gorge level. This ramp represents centuries of weathering and small rockfalls that have helped wear back the cliff.

> Learn more about inter-reef areas in chapter 3, pp. 76–77, and vugs in chapter 8, pp. 152–53.

Climbing back up to the rim of the main gorge you enter heavily treed areas above the Icicle Walls (*Site 17*). Active and dry tributaries enter from the north (Middlebrook) side. One stream makes its way through an impressive series of terraces in rock to become a **hanging valley** that drops sharply down to the main Grand.

A little way upstream, halfway to the Irvine Junction, tumbled terrain records where one or more former tributaries used to enter the Grand (*Site 18*). Now they are dry valleys. The gorge wall on this side is divided into several irregular towers. They rise to the level of the main cliff and overlook deep depressions filled with trees, leading into small caves. It is an ideal place to scramble around and note the fossils, cliffs at different levels, and varied erosional landforms. At least, my children used to think so, and named the area "the Castles." Each child claimed a particular knoll and depression as her castle and dungeon, whether defending against others or inviting them to a picnic. Among the Castles is the largest coral colony I have encountered in the gorge, whose secret I keep.

Excursion 3: Lower Circuit of the Grand

This third excursion takes in the lowermost Grand Gorge, between the Middle and Low Bridges in the conservation area (fig. 2.19). As noted above, the itinerary begins and ends at the Middle Bridge (*Site 15*). The starting point and some of the

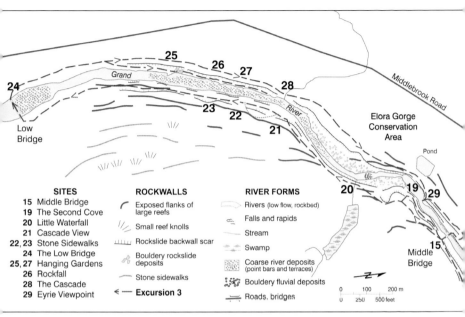

Figure 2.19 **Excursion 3: Features and sites along the Grand River between Middle and Low Bridges.**

numbered sites are accessible along the park roads. The whole route first follows the left bank of the river and returns along the right via the "Low Bridge" (*Site 24*) where the gorge ends.

This excursion includes the most spacious and majestic sections of the whole gorge. Below Middle Bridge the bedrock outcrops are more extensive and create larger spaces and rock walls. Views toward the right bank slopes seem to climb much further northward than parts visited so far.

Immediately downstream of Middle Bridge are a narrow slot and cataract, below which the river enters and expands into a generous open space known as "The Second Cove" (fig. 2.20 and *Site 19*). The smooth surface of this expanse of normally quiet water reflects the cliffs of the far side. Plenty of debris

Figure 2.20   **Entering the Second Cove** *(Site 19)* below Middle Bridge.

is swept through the steep, narrow slot under Middle Bridge. However, except in the highest flows the debris is stalled by the gentler, wider channel further down and a sudden drop in stream energy. In the Second Cove, the river swings sharply from the left to the right side, and back again. It is deflected by the largest boulder deposits or bars yet encountered. Below the Second Cove the river increasingly resembles and behaves like the alluvial streams otherwise absent from the gorge.

The left flank of the gorge is complicated by a series of rocky terraces and old cliff lines. They crowd the river and create steps upward and away from the water—a continuation of features described in the quotation from John Connon that starts this chapter. The walls are similar to, but larger than, those described between *Sites 10* and *14* on Excursion 2. Along with the exceptional cliff above Middle Bridge, on the right side the bedrock steps create a general westward migration and deepening of the gorge. This helps explain differences between the right and left banks, noticeably features hidden in the woods around them.

The old cliffs on the left side do not intersect the permanent water table. Springs and seepage are rare. Almost everywhere on the right flank, however, there is a single vertical or over hanging wall, and many springs and precipitated aprons of **tufa** (a freshwater **carbonate** that is an important source of karst features in limestone terrain). There are fewer signs of the old river levels on the right side, mostly narrow and lost in denser woodland.

Learn more about tufa in chapter 8, pp. 159–63.

Continuing down the left flank, you encounter a hanging tributary (*Site 20*). It drains what used to be a small artificial lake in the conservation area, now filled in, leaving only a small stream and wetland. At the rim of the gorge is a small notch where the stream plunges down through rockfall boulders and dead tree trunks, then descends to the gorge floor as a true waterfall.

About 150 metres beyond this little waterfall a dry valley descends in a series of steps parallel to and well above the Grand. This cut turns sharply to the right and provides an easy

Figure 2.21   The Cascade (*Site 28*), near the lower end of the Grand Gorge in the conservation area, seen from *Site 21*.

descent to river level. The viewpoint is *Site 21*. The Cascade tumbles down the cliff opposite, at *Site 28* (fig. 2.21). A curtain of white water more than twenty metres high splits into skeins and rivulets as it falls to the river. It is fed by another small, hanging tributary from the Middlebrook area.

Learn more about hanging tributaries in chapter 6, pp. 126–29.

From here you can, if you wish, follow the riverbank. You will find that the gorge begins to wane, its channel less steep. Reeds and bushes colonize the valley floor and help to trap the sediment coming through. Above the left bank of the river you encounter a tumbled terrain of abandoned cliffs and terraces, evidence of former river levels. There are strath terraces across the slope, roughly parallel with the river but set back from it.

Figure 2.22   **Example of "Stone Sidewalks."**

On this left flank of the gorge, you will find cliffs set back from the river and offset from one another. The paths tend to be sheltered beneath hanging rock. Here, too, you will find some of the most impressive examples of the intriguing forms known as the **"Stone Sidewalks"** (fig. 2.22 and *Sites 22–23*). They usually mark the first and second main benches above the river. Higher up, the country between the cliff lines and the gorge rim is mostly wooded. From the trees at the river's edge you have excellent views of "The Cascade" (fig. 2.23).

Learn more about hanging rock in chapter 7, pp. 133–39; and about the Stone Sidewalks in the same chapter, pp. 139–41.

Figure 2.23   **"The Cascade"** *(Site 28)* in early spring.

Down valley and southward, the Sidewalks gradually subside, until disappearing where the Low Bridge sits (*Site 24*). Here the gorge is mostly defined by a single main cliff, rising abruptly from the river. A little further downstream, cliffs and outcrops of dolostone terminate altogether. Soon they are buried under glacial and alluvial deposits, while the river opens into the wide, flower-strewn flats below Inverhaugh and flows down to embrace the generous flood plain of Wilson Flats.

Excursion 3 continues across the Low Bridge (*Site 24*) and returns along the right bank of the Grand. The Low Bridge was damaged by a severe flood in 2017 and had been reopened to pedestrians at press time but was still closed to vehicles—please check with the GRCA for current conditions. On this side, the highest terrain continues above the rim of the gorge. At the base of the cliffs there is much more moisture than in other areas discussed in this book. Where there are small ledges and depressions, the ground is soggy. The trees cluster close together, in places almost like rainforest. Such conditions are difficult to navigate, making it more tempting to proceed along the conservation area road that lies above and runs parallel to the river.

> Learn more about the 2017 flood and other extreme weather events that have affected the gorge in chapter 6, pp. 116–20.

Behind trees and clinging to the steep slope along this part are cones, small walls, and terraces of tufa, from which emerge springs of lime-rich groundwater. Tucked in along the base of the cliffs, starting from a little way upstream of the Low Bridge, are springs that feed the calcified aprons of the "Hanging Gardens" (fig. 2.24). They continue upstream almost to the Second Cove (*Site 19*). These living **travertine** aprons can be visited at *Sites 25* and *27*. You can reach them with difficulty by descending from above, but if you take the path under the cliff, they are almost continually at your side.

Figure 2.24 **Example of "Hanging Gardens."**

Learn more about "Hanging Gardens" in chapter 8, pp. 159–63.

Down at river level are numerous rockfalls that have descended from the sheer cliffs of the right (west) bank. A large and fairly recent example is found at *Site 26*, reflecting at least two events, the latest of which occurred in 1988, when large slabs fell from the flank and underside of an overhanging reef mass. The broken rock smashed through the trees and undergrowth to reach the river.

Figure 2.25
Looking down into the gorge and part of the Second Cove *(Site 19)* from "Eyrie Viewpoint" *(Site 29).*

Learn more about rockfalls in chapter 7, pp. 141–44.

Near *Site 25* you may prefer to return to the gorge rim. You can follow narrow, wooded paths along the cliff edge to get closeup views of the head of the Cascade (*Site 28*). This large drop is another place to think about the history of the Grand, and speculate on why the main gorge is so deeply incised.

At the end of Excursion 3, as you near the Middle Bridge, one last impressive prospect opens up. The gorge rim culminates in a fenced-off cliff at "Eyrie Viewpoint" (*Site 29* and fig. 2.25), where you can look directly down on the Second Cove (*Site 19*). This cliff has the greatest vertical span in the gorge. At the extreme tip, the promontory overhangs and closes off the upstream rockface. A hardy grove of cedar trees clings to the rock and adds to the height.

VISIBLE AND HIDDEN IN THE LANDSCAPE

The excursions in this chapter are intended to introduce the kinds of landforms and developments encountered in the gorge, concentrating on what can be seen in the landscape. The later chapters return repeatedly to these visible forms. However, the interpretation of landscapes demands attention to much that is not directly visible, some not at all. Earth science recognizes the importance of processes on scales too large, or too small, for direct human observation. Hidden realities are, especially, those involving time and history. After all, landforms comprise the solid Earth surface. Being composed of solids, they can resist change and persist for centuries, millennia, or far greater spans. They take us into the deep history of the Earth, and major changes in the distributions of continents and oceans.

Figure 3.1
Bedrock supports and
helps shape all major gorge
landforms: caves and
buttresses, river channel and
cliffs, notably hanging rock.

Chapter Three
Bedrock

"to those who love to study nature the rocks have a wonderful story to tell."
—John Robert Connon (1930)[1]

The gorge landscapes record the work of Earth surface processes, mainly interactions of stone, streams, and gravity. Bedrock is the most important material. It consists entirely of a type of limestone known as dolostone and is part of a single geological time horizon, the Guelph Formation. The bedrock's story began in an area of tropical seas where buildups of limestone continued over many millions of years. This chapter explores the conditions that brought it into existence, and situates the gorge dolostone and fossils within the broader picture of geological history.

———————

IN THE SILURIAN

The Guelph Formation was named in early surveys around the city of Guelph, southeast of Elora. It appeared as such in William Logan's classic *Geology of Canada* (1863).[2] The formation

is encountered in a narrow band of outcrops west of the Niagara Escarpment, and from the Bruce Peninsula to Niagara Falls
(fig. 3.2). Most of what is mapped as Guelph Formation rock is
buried under much later, superficial sediments. Exposures tend
to be modest. Those in the gorge are among the more notable in
lateral and vertical extent. Where the gorge walls are highest,
at "Eyrie Viewpoint" (*Site 29*), a continuous Guelph sequence
is exposed up to forty-five metres thick. A rock core drilled at
Elora identified ninety metres of the Guelph Formation near

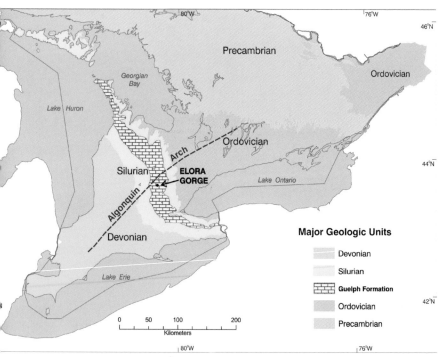

Figure 3.2   **The major bedrock subdivisions of Southwestern Ontario, highlighting the
Guelph Formation and the Algonquin Arch.** Modified after Ontario Geological Survey,
Bedrock Geology of Ontario, Southern Sheet. 1991. 1:1,000,000 scale, Map 2544, http://
www.geologyontario.mndm.gov.on.ca/mndmfiles/pub/data/records/M2544.html.© King's
Printer for Ontario, 1991. Adapted and reproduced with permission.

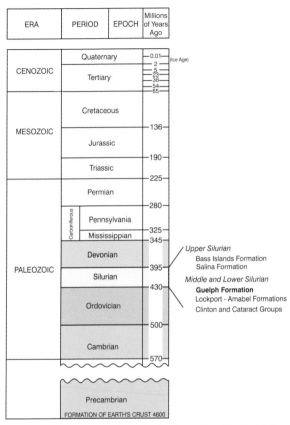

Figure 3.3.   **The geological column showing the position of the Guelph Formation in the global Silurian Period, and the Silurian in the Paleozoic.**

the village. About 120 metres can be discerned of the total vertical profile, but neither the very earliest horizons, nor the latest, have been identified in the gorge.

The gorge rocks can be situated within the larger scheme of Earth history, beginning with their place in the geological column (fig. 3.3). They mark the last phase of the Middle *Silurian*

Period, named after an ancient Welsh people, and indicating the early influence of British geologists. In turn, the Silurian is part of the global *Paleozoic Era*.[3] "Paleozoic" means "old life" and was formerly thought to encompass the earliest stages of life on Earth.

The Paleozoic involved a series of gigantic expansions of life forms, and several major extinction events. It began with the greatest single expansion in the extent and variety of life forms on Earth, the "Cambrian Explosion." That was roughly between 541 and 530 million years ago. The Silurian itself is placed between 444 and 419 million years ago. It also began with a rapid, widespread expansion of life forms. The Guelph Formation spanned some 5–7 million years, from 427 to 423 million years ago. The Paleozoic ended around 250 million years ago with yet another extraordinary mass extinction, and some of the largest losses of life on record. They included nearly all the species found as gorge fossils.

The seeming great age of the gorge rocks does not mean they are anything like the earliest in Ontario. Some kilometres north of Elora are surface rocks dated up to 2.5 billion years old.[4] Recent estimates place the origins of life on Earth as much as 3.7 billion years ago.

REEF LIMESTONES

"Corals in considerable variety may be picked up almost anywhere, and singular as it may appear, even a sponge."
—David Boyle (1875)[5]

The key concern of this chapter is rock control over landforms. This depends on bedrock constituents. In the case of sedimentary bedrock such as that of the Guelph Formation, these constituents can include the original material laid down and compositional change due to geothermal heating, migrating fluids, or metamorphism. Also, there can be configurational change such as folding and faulting by earth movements. In sedimentary rocks the influence of grain size can be critical, with important differences in response between sand and silt, boulders versus clay, and the juxtaposition of different strata, such as sandstone versus shale. A full range of grain size types can be observed in the gorge rocks.

The gorge limestone raises two key issues: first, the bedrock was laid down in a setting vastly different from its present location—in a tropical marine environment. Second, it is a unique product of living organisms. As with most limestones, the rock in the gorge came about through conversion and fixing of dissolved and gaseous elements, mainly by marine organisms.

The ingredients of the gorge limestone were derived from sunlight and waters in the marine basins where they were deposited. *Calcium carbonate* was the main product. A preferred term for limestone is carbonate, which signals the paramount role of carbon. Limestones comprise huge stores of carbon.

In the Silurian seas, settings were associated with coastal, reef margin, reef ramp, lagoonal, and offshore conditions. The reef-building organisms had to be able to resist tidal currents and waves, although they could be damaged in the fiercest storms. Their rubble might be thrown into a lagoon, or to the sea floor. The carbonate muds around the reefs were home to organisms that burrowed through the sediment, or filtered the grains, adding further complications.[6]

Life forms were, and their remains are, critical in the distribution, volume, and geometries of the Guelph limestone. They were intimately involved in dissolving, fixing, and accumulating carbonate, and in protecting it from erosion by wind and waves. Most important were creatures attached in place, or **sessile organisms**, notably corals and, above all, those that built and inhabited organic reefs. Different species or colonies, and combinations of these marine organisms, were responsible for the mass and layout of reef forms.

Another controlling influence was fluctuation in sea level, which changed habitat conditions for reef-building organisms. The depth of sunlight penetration is always critical. Reefs cannot form in the perpetual gloom at great depths. Sea floor topography is hence crucial to where they flourish, or not. Meanwhile, reef-building itself modified sea floor topography and local elevations. To an overwhelming extent, however, the gorge rock reflects the varieties of marine organisms present and their activity, primarily reef growth and ecology. The exposures of the Guelph Formation from Wellington County and the city of Waterloo (around twenty kilometres south of Elora) northward have been classed as a barrier reef complex.[7] Its development resembled today's scattered reef archipelagoes of the Bahamas, and in some respects, the Great Barrier Reef of Australia.[8]

When you look down into the gorge from the various viewpoints, or when you walk or float along the Grand or Irvine Gorges, in effect you are exploring the anatomy of the Guelph barrier reef. You are negotiating a labyrinth of reef colonies and other features reflecting carbonate deposition or patterns of its resistance to erosion. The gorge terrain serves to pick out the forms, dimensions, and legacies of organic growth left by the changing environments in the prehistoric Guelph Sea. Where the rock is exposed, your boots crunch their way through a graveyard of countless Silurian fossils.

LIFE AND LANDFORMS

The gorge bedrock originated with inundations by tropical seas and buildup of limestone around their shores. It introduces a theme that recurs throughout the story of the gorge: the interdependence of landforms and the living world.

Geomorphology tends to emphasize physical processes: land surface processes and geometry, gravity, geological structures, and tectonics. These are very important, of course. But Elora Gorge is a reminder of the often equal or greater roles of the living world. It would be difficult to overstate the barrier reef environments' legacy for what one sees today, or the enduring influence of reef organisms.

There are debates about just what a reef can consist of, and how much ancient examples were like modern ones. In Earth science, "reefs" refer strictly to forms that are biologically constructed. Today's large reefs, including the most familiar type, coral reefs, flourish in tropical oceans. Sea surface temperatures must be high, normally between 25 and 30 degrees Celsius. The conditions for reef-building were probably similar in the Guelph time. However, global sea levels were much higher than they are today. Vast continental areas were drowned in relatively shallow seas called *epicontinental* or *epeiric*. They covered most of the nascent North American continent and gave rise to vast limestone areas. The Guelph Formation is a relatively minor but striking example. It is important to be aware of how various organisms contributed to the mass and geometry of the barrier reef, and what links them to the forms left behind. Present-day landforms reflect ecological conditions or organic growth forms of the barrier reef and of intra-reef and coastal environments. The shellfish and corals, their shapes and colonies, influence smaller scale, distinctive units.

As noted earlier, the life forms in the Guelph Sea became extinct long ago but their fossils record a rich sea floor ecology. More than 150 species have been identified in the gorge rocks. In places a fossil hunter can be well rewarded. Plant life must have been abundant, serving, as always, to feed most other creatures. Mats of algae likely coated much of the lagoon areas and reef flanks. Their remains are, however, hard to reconstruct, while the easiest are fossil shellfish.

The fossils in the gorge are entirely of marine life belonging, as noted above, to the Silurian Period and a single formation, the Guelph of the mid-lower Silurian. To search these out, a magnifying glass and a good pocket guide to fossils will help.

In general, barrier reef sites can exhibit a great diversity of species and colonies. Candace Brintnell's description of a portion of the Guelph Formation on the Lower Irvine Creek gives a rich sense of this diversity:

> The megafauna are the largest in size and most abundant of all three areas [examined], comprising *stromatoporoids*, common fragmented *bryozoans* and sparse corals: *Favosites, Halysites,* and colonial *rugosans. Stromatoporoids* consist of both laminar and domical types, ranging

Figure 3.4   **Cliffs, buttresses, hanging rock, and riverbed illustrate how bedrock supports and helps shape all major gorge landforms.**

in size from 5–15 cm in diameter for domical shapes; and 5–20 cm in length for laminar forms. The small *Favosites* are golf-ball-sized and other corals are smaller. Small microbial-*stromatoporoid* mounds are common at the base of this unit, ranging in size from 1–6 m. Shell debris is scattered throughout and rare conical *gastropod* shells are locally present.[9]

This passage introduces and illustrates the variety of fossils identified in the Guelph Formation and present in the gorge. "Selected Guelph Formation fossils" in chapter 9, page 174, lists a selection of Silurian fossils named after sites, places, and personalities associated with Elora Gorge, including David Boyle. However, the foremost interest and the critical link to the landforms are the reef-building organisms. Their massive constructions prefigure the more impressive landforms of the gorge walls and geometry of river channels (fig. 3.4).

REEF BUILDERS

The rigid architecture of the reefs was erected mainly by, and identifies, the *frame builders* or *constructors*. Some were ancient species of coral, whose colonies could reach up to a metre in diameter. Corals, *Anthozoa* of the phylum *Cnidaria*, are found throughout the gorge, but are generally small and easily missed. Many are poorly preserved, but some are the most attractive fossils, including "honeycomb" and "chain" corals (see fig. 3.5 for illustrations). A few can form larger units: I know several *Halysites* colonies more than half a metre across, but most coral colonies were smaller. And they were not the main frame builders in the gorge rocks. That was the work of another class of invertebrates, ecologically close to the corals but belonging to the sponges, phylum *Porifera*.[10]

Various types of sponge remain important in today's reefs.

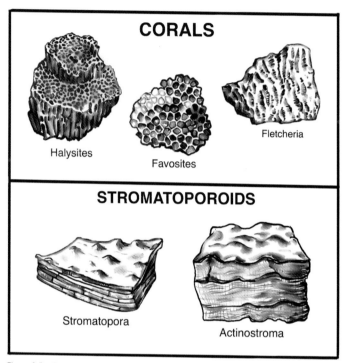

Figure 3.5 **Examples of the stromatoporoids and corals of the Guelph Formation that can be encountered in the gorge.**

Those in the gorge rocks became extinct long ago. They are known only by their scientific name, the *Stromatoporoidea*.[11] Some pronounce it "stromatoPORoids," others "stromaTOPoroids." Both are a mouthful. A short form will suffice here: "Stroms." The literal translation of *stromatoporoid*, "porous layered," conveys what is typically visible. Even more literally, one might compare them to a bedspread full of holes. In a pioneering study of Ontario examples, W.A. Parks in 1907 referred to "the puzzling and interesting group of the Stromatoporoids" and remarked on the difficulty of distinguishing different species without scientific equipment.[12] While most are not well preserved, in places along the gorge intriguing details survive in some of the most massive and striking exposures of the dolostone (see fig. 3.5 for illustrations).

In most cases what you see are the crudest moulds, odd slabs or dumplings of stone; something that might suggest a cut-away section of a calcified cabbage (fig. 3.6). Smaller colonies etched out on vertical rock walls, can resemble the rings of an onion. In a few places the naked eye can make out well-preserved colonies or their remnants, from a few centimetres across to several metres. Some reveal intricate forms similar to crochet work. Most were spoiled in the process of forming dolomite, which we will explore a little later in this chapter.

Stroms performed a variety of functions in the reef systems and in shaping features typical of the dolostone bedrock. They

Figure 3.6   **Section through stromatoporoid colonies in the Elora Gorge.**

created and inhabited hard **calcite** or **aragonite** skeletons. Typically, they are found as compact *domal* colonies (that is, having the form of a dome) (fig. 3.7). Some grew as many-branched, tree-like structures; others as mats or mattress-like slabs. It is hard to overestimate their roles as the principal reef builders. Major remnants of their colonies literally stand out in the landscape. They comprise the most salient and resistant forms in the gorge. Some survive as narrow towers that mainly grew vertically. Others grew as much laterally, and some created extensive horizontal, tabular beds.

The Stroms did not act alone and, important as reef builders were, other creatures performed special functions, including reef-dwellers. They deserve attention too. The *binders*, as the term suggests, served to fix and bind in place other, more open, porous or weaker growths and sediments. Binders added to the mass, density, and stability of the reefs. They included some species of Strom, corals, and algal mats. *Bryozoans*—meaning "moss creatures"—grew on and over other organisms, much like mosses on boulders and rainforest trees. Another group, called the *bafflers*, were critical in near-surface parts of the reefs affected by waves and currents. As the term suggests, they helped reduce, slow, or block water movement, tending to protect and preserve reef mass. Some corals performed this function.

At low tide, rocky islets of porous white rock may have emerged, to be washed over by the waves. But no seagull left its telltale droppings on the rock. Birds were not yet part of the living world. In the blue depths of the water, looming outlines of reefs appeared and hints of the life that teemed about them. Most important were the reefs themselves, constructions that prefigured the more impressive landforms of the gorge walls: the towers, overhangs, and otherwise massive and diverse geometry of river channels (fig. 3.4).

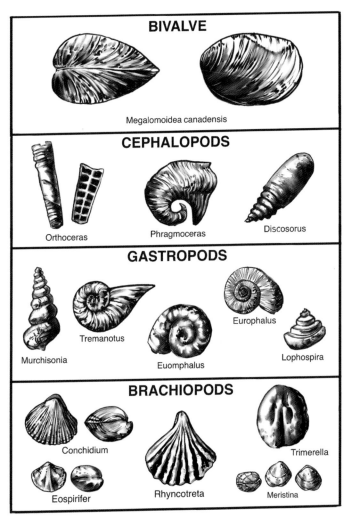

Figure 3.7   **Fossil shellfish *(Mollusca* and *Brachiopoda)* examples of which can be found in Elora Gorge.**

It is fascinating to consider how present-day landforms reflect ecological conditions and organic growth forms of the barrier reef, intra-reef, and coastal environments. The shell-fish and corals, their shapes and colonies, influence smaller scale, distinctive units. But one can hardly ignore the ancient shellfish, to which we now turn.

THE SHELLS

Among the fossil shellfish you will find long-extinct relatives of oysters and clams—known as *bivalves* to *paleontology,* the science of ancient life. If there were space travellers 400 million years ago, the Guelph Sea and barrier reef would have been a rewarding stop for seafood. Clam bakes would surely have been high on the list! Visitors would also have found ancient relatives of today's edible whelks, sea snails, periwinkles, and conches; members, that is, of the *Gastropods.* There were fore-bears of the *Cephalopods,* cuttlefish, and the elegant *nautilids.* These are all members of the great life group or *phylum* of the *Molluscs.* They take their name from the Latin *mollis,* meaning "soft-bodied." What you find in the gorge however, are their original hard parts, the shells (referred to as *valves*), or the moulds and casts in the rock that formed around them (see fig. 3.7, middle rows, for illustrations).

Other typical shellfish of the Guelph Formation belong to the phylum *Brachiopods.* Expect to find examples as small as your little fingernail and up to the size of a hen's egg. Most of more than thirty species known in the Guelph Formation in Ontario can be found in the gorge (fig. 3.6, bottom row). They include the lamp shells, which preferred shallow and sheltered waters where they could live burrowed into the **calcareous** mud that settled around the reefs. You can sometimes see lamp shells lined up along, or just below, a break in the strata or between successive reefs, close to the water surface at that time.

Megalomoidea canadensis is among the more readily discovered and numerous shelled species, and the largest (fig. 3.7,
top row). An extinct relative of today's clams, it used to be called
Megalomus canadensis. The shell form, with its two heart-
shaped valves and ribbed exterior, is sometimes clearly visible.
More often, however, it is poorly preserved. You might see a
mould of the body like a large stone plum or, elsewhere, a series
of airfoil-shaped hieroglyphics etched into a rock wall. These
clams favoured sheltered lagoons or *inter-reef areas* where, if
you discover one, you will likely find hundreds. And they may
be found in a wide variety of settings within the barrier reef. In
the walls of the gorge there are extensive clam beds with countless individual shells; for instance, just beneath David Street
Bridge on the Irvine (*Site 5*). Readily observed, this clam was
used as a marker of the Guelph Formation. It appeared shortly
before the Guelph time and went extinct at its end.

> Find fossil corals, Stroms, and shellfish on Excursion 1,
> Sites *4–5* and Excursion 2, Sites *12–14* and *18.*

The clams may be among the larger, more readily discovered fossils, but perhaps lack the elegance of some others. The
coiled shells of the *Gastropods,* introduced earlier, can be
elegant and handsome (fig. 3.7, third row). Over sixty of their
species have been identified in the Guelph Formation. Most
are between one and three centimetres in diameter. You may
more easily catch sight of them where the rock has fractured
along the outer surface of the shell to reveal its spiral form. In
some cases the growth lines can be distinguished, dividing
up the rounded whorls. Some shells are squat, some globose,
a few high-spired. Among the latter is *Murchisonia boylei,*
named after both David Boyle, the Elora schoolteacher, and
the notable British geologist Roderick Murchison, who named
and led the worldwide adoption of the Silurian Period.

DOLOSTONE

After the reef materials were laid down, further changes occurred, from living reef to lifeless rock, and effects of settling and compaction, ultimately expelling most of the sea water. The gorge rocks were buried under later formations hundreds of metres thick. Most important were effects that led to dolostone.

The carbonate materials originally laid down were, as noted, *calcitic*, mainly *calcium carbonate*. Later, there was a substantial chemical change, replacement of a large fraction of calcium with magnesium ions. The resulting mineral is dolomite, with roughly equal amounts of calcium and magnesium.[13] *Dolomitization* also involves pervasive recrystallization.[14] In the process the rock becomes denser and more cemented. The fossils are often spoiled.

The end of the Guelph time was marked by local and widespread fall of sea level, notably where the limestones of Southern Ontario had been laid down. The rocks immediately overlying the Guelph comprise the *Salina Group*. As the name suggests, this group marks an increasingly saline marine environment. It is noted for substantial salt beds and evaporites including deposits of *gypsum*. The barrier reef was replaced by salty lagoonal and coastal conditions. To the west and south, Salina rocks are exploited by salt mines, as at Goderich and Windsor, Ontario. They give rise to saline groundwater aquifers but are exposed rarely at the surface and not at all in the gorge area. It is reasonable to suppose they once covered the Guelph Formation here but have been completely removed by erosion or in solution.

THE FIRST AGE OF THE GORGE

The time when the Guelph Formation was being laid down can be called the *First Age* of Elora Gorge, identifying its initial place in Earth history. Specifically, the barrier reef defines the conditions under which the gorge rocks formed. Challenging questions arise, however, about world geography and how much it has changed since then. Why, for example, are tropical reefs and marine fossils found today under snowy, near-boreal conditions in the heart of the continent? To answer these questions requires a wider sense of geological history, a shift in perspective and concerns. And it turns upon a revolutionary change in what Earth science contributes to landscape interpretation.

Figure 4.1   The detailed textures, small and large cavernous weathering, and variable cliff faces all speak to the distant origins of the Guelph dolostone, but here too is a remarkable story of earth movements and transports over 400 million years.

Chapter Four

Tectonics and Ancient Geographies

After the gorge rocks were laid down, their fate became linked to major changes in the Earth's continents and oceans. Before considering these changes, however, it is important to appreciate certain persistent relations to geological structures in Southwest Ontario. These existed when the gorge rocks were being laid down and endure alongside them to the present day.

SLEEPING GIANTS

In the Silurian Period, as now, the upper Grand area lay at the intersection of two major structural features, the *Michigan Basin* (fig. 4.2) and the *Algonquin Arch* (see fig. 3.2 in chapter 3, page 62). These overlap in the vicinity of the Elora Gorge. In the early history of the gorge rocks, Earth movements affecting the two structures were important controls on where and how the barrier reef developed. Episodes of subsidence of the Michigan Basin and uplift of the Algonquin Arch affected details of sea level, the pattern of deep and shallow areas, and coastal features.

The Michigan Basin is a nearly circular feature, identifiable today in the concentric bedrock outcrops that dip toward the

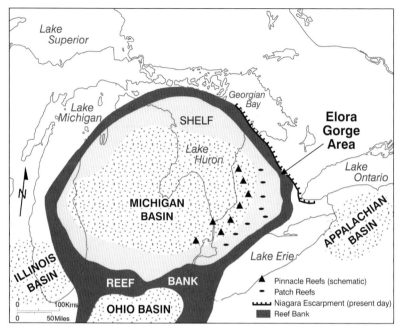

Figure 4.2   Schematic sketch of the Michigan Basin in the Guelph time indicating structural, reef, and depositional features southwest of the gorge area (after Briggs and Briggs, 1974, 2, Fig. 1).

basin centre. The gorge rocks occur near the rim, within an almost continuous sequence of marine deposits. The gentle dip of the gorge rocks toward the southwest marks them as part of the basin. Near its outer, eastern rim, the barrier reef formed an almost continuous belt. A shallow marine platform developed and responded to the interplay of earth movements and sea depths.

It is estimated that the Michigan Basin was tectonically active from about 530 to 340 million years ago, including all of the Silurian Period. Episodes of subsidence are recorded in the series of marine sediments that eventually filled the basin.

The geological cross-section shows the gorge rocks to occur within an almost continuous sequence of marine deposits. These reached more than 4,000 metres in thickness at the basin centre. The Guelph time came almost exactly at the mid-point of the basin and arch activity. It is thought that more than twenty, possibly as many as thirty, *marine transgressions* occurred in the Paleozoic Era.[1] These were geologic events during which sea level rose relative to the land and shorelines moved toward higher ground.

The Algonquin Arch is an ancient ridge in the underlying Precambrian rocks. It trends northeast to southwest, midway between Georgian Bay and Lake Ontario. The Elora Gorge area sits astride the Arch, which exercised several important functions, including control of sea level and depths in the immediate area. Repeated small uplifts maintained a some-what raised platform, where the reefs flourished. A marked thickening of the Guelph Formation in the upper Grand seems to record more favourable reef-building conditions there.

The alignment of the Arch prevented sediment being car-ried from the northeast toward what is now Southern Ontario. Further east, at that time, highlands, mountainous islands, and volcanoes were being uplifted and rapidly eroding. Muddy seas could have smothered and prevented reef growth. They affected the local environment before the Guelph time and would do so afterward. For the moment, however, the Algonquin Arch helped keep the seas clear. It also helps define the southeast margins of the Michigan Basin, which is of similar age.

While the region was spared major tectonic upheavals, small Earth movements affected and helped control the his-tory of the reefs. The tectonic activity was, in turn, controlled by what was happening at depth in the Canadian Shield. It would also continue to define and constrain the layout of Southwestern Ontario's bedrock into the present day. How-ever, folding of the Guelph Formation rocks is negligible. The original rock layers remain nearly horizontal. There is a slight dip toward the southwest, but it is not readily apparent on the ground (fig. 4.3).[2] Geological faults, which can divide up and displace bedrock, are few. Those that exist in the surrounding bedrock tend to die out in the gorges area, or to exert only minor influence.

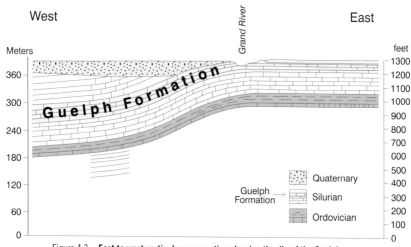

Figure 4.3   **East to west vertical cross-section showing the dip of the Guelph Formation's bedrock in relation to the Grand River.** Modified after M.Y. Williams, *The Silurian Geology and Faunas of Ontario Peninsula, and Manitoulin and Adjacent Islands*, Memoir 111, Geological Series No. 91 (Ottawa, ON: Geological Survey of Canada, 1919), unpaginated insert entitled "Structural Section along Line A-B," https://archive.org/details/cihm_82237. Reproduced with the permission of the Department of Natural Resources, 2022.

Thus, for most of their geological history, the Michigan Basin and Algonquin Arch have been like "sleeping giants." On the one hand, the Canadian Shield ensured more stability than change to the gorge rocks. Since the latter were laid down in the Silurian, they have remained in almost the same relation to the two larger structures, an enduring structural constraint within a larger story of movement and change. On the other hand, this constraint seems to contrast with, but gives greater prominence to, surface changes in regional geography.

THE GREAT JOURNEY

Since their formation, the gorge rocks have been caught up in a huge, meandering journey across the planet's surface. This journey dominates the *Second Age* of the gorge. Understanding what happened requires a shift in perspective, moving from a focus on the visible scene to broader scales of time and geography. Key developments are not directly observable in the gorge. The ideas that would inform the new paradigm had barely entered scientific thought before the late twentieth century, and have only been accepted in the past few decades. That is, a century and a half after the founding of Elora. And they still drive multiple research questions across the Earth sciences.

Through this roughly 400-million-year phase—the longest part of gorge history—the regional bedrock and local geography went largely unchanged, thanks mainly to the "sleeping giants," the Michigan Basin and Algonquin Arch. Meanwhile, however, the gorge rocks occupied and travelled through a succession of widely separated places in world geography. They were part of an ever-changing configuration of continents and oceans.

Toward an Explanation: Plate Tectonics

There were early attempts to explain how, for example, Silurian corals are found in high latitudes. The more influential early ideas involved the concept of *continental drift*, quite widely supported before the Second World War. Global geography was attributed to the splitting up of former continents, and their remnants drifting together, or apart. For instance, coastlines and geological formations on opposing sides of the Atlantic Ocean seemed to fit together. However, by the 1940s continental drift had been dismissed. Influential geoscientists decided that it was physically impossible for continents to drift.

For a time, efforts to explain corals in polar regions and an ancient ice age in the Sahara Desert turned to climate change.

Shifts in the Earth's orbit were invoked, increased solar or volcanic activity affecting the atmosphere. Some argued that these could bring drastically different north-south arrangements of heat and cold, deserts and snowfall. However, despite a growing conviction about the importance of climate change in other landform developments, it proved inadequate to explain the geography of continents and oceans.

By the 1960s, greater refinement of surveying and satellite measurements revealed that the continents *are* indeed moving relative to one another. The rates of travel are generally very slow, centimetres annually, but real enough. And they are adequate to drive major changes over geological timeframes. What was needed to explain these movements was a mechanism.

A revolutionary change came with the discovery of *ocean floor spreading*.[3] The mid-ocean ridge systems proved to be fed by convection currents in basaltic rock that wells up from below, constantly bringing new mass to the surface. In turn, the raised mid-ocean floor spreads or slides outward from the mid-ocean ridges along gigantic faults. Large-scale shifts in thermal and gravitational conditions add to the forces at play. The continents are carried along by processes that occur deep below the Earth's surface.

Mid-ocean ridges turn out to define major boundaries between the Earth's crustal units; the *plates* of *plate tectonics*.[4] These engage the more mobile *mantle*, the solid Earth between the crust and core. As they spread outward and away from mid-ocean ridges, opposing plates meet and can create highly dynamic zones, notably **collision zones**. Mountain belts may form, or deep ocean trenches where rock mass descends back into the crust. Plate tectonics provide a comprehensive theory of the Earth's development. It draws together understanding of volcanic activity, earthquakes, the shape of the oceans, and mountain-building forces.

It was thanks to these forces that, since the Guelph time, the gorge rocks in effect "went walkabout." In their great journey, they were recording plate tectonics and related climatic and surface changes, not least how these affected the fate of the barrier reef.

Plate tectonics reveal that the layout of land and oceans at the start of this journey was radically different from today. At the beginning of their displacement, the gorge rocks lay amid unfamiliar lands and vast marine features; a tiny patch in the Southern Hemisphere tropics. While there was no Atlantic Ocean, global sea levels were much higher than today. The topography of the continents was generally lower and more subdued. Rather than several oceans there was, essentially, just one; a vast area of saltwater occupying most of the northern hemisphere. It has been named *Panthalassa*, meaning "All Sea." For most of the Paleozoic, land masses were pushed together into a single "supercontinent" named *Gondwanaland*.[5]

In the Doldrums

According to recent work, during the Silurian the gorge rocks would have been in the southern hemisphere. There was no Atlantic Ocean. The Himalayan Mountain range did not exist! The Sahara sat over the South Pole, where the waning part of another ancient ice age was found!

Recent reconstructions place the gorge area somewhere between 15 and 25 degrees *south* of the Equator, in latitudes known as the *doldrums* (or more technically, the *intertropical convergence zone*). At that time and location, the night sky would have been filled with stars now seen from, say, Fiji and Tonga in the Pacific Ocean.

The doldrums are regions of variable and uncertain winds, once notorious for becalming sailing ships. At noon, throughout the year, the sun is high overhead. On cloudless days it shines fiercely. A still and oppressive heat may last for weeks. In the Guelph time the evidence points to similar conditions, but the spin of a younger Earth was faster. A more dazzling sun followed a speedier course. Apparently, a whole day lasted barely twenty hours. The heat could be broken by sudden storms that brought waterspouts and torrential rains. Tropical cyclones, hurricanes, and typhoons, with their associated storm surges, still achieve their greatest intensity here. There is evidence for these from the time when the gorge rocks were forming.

After the Silurian, an even greater concentration of continental areas developed: a supercontinent known as *Pangaea* ("All Earth"). It reached its maximum extent around 200 million years ago. The gorge rocks had travelled to roughly three degrees north latitude by then, near the contact between the North American and African plates. Thereafter, Pangaea started to break apart, a process that would come to define the fate of Earth geography down to the present time.[6] Eventually, a recognizable configuration of the Earth's surface began to emerge with an Atlantic Ocean and a more familiar placement of the North American continent and gorge rocks. This journey was slow by human measure —on average about the same speed as hair or fingernail growth—and did not proceed at a constant rate. Moreover, the direction of movement followed meandering and sometimes sharply redirected courses.[7]

The North American Craton

Plate tectonics have transformed Earth science. But here again, the gorge raises special and less widely mentioned developments. In places like Southern Ontario the more dramatic outcomes are less evident. Throughout time, the location of the gorge rocks has never coincided with high mountains, mid-ocean ridge systems, volcanic arcs, or highly active earthquake zones. Rather, the Guelph rocks are part of other developments that have also affected continents. Once thought merely to drift, the continental areas of plates have also hosted immense buildups of sedimentary rock across their surfaces. Like the Michigan Basin, these are mainly legacies of formerly extensive epicontinental seas.[8] Locally they may appear modest compared to mountain buildups, but they affect vast tracts and have key, albeit relatively neglected, roles in plate tectonics.

The Michigan Basin illustrates such a link between the Guelph Formation rocks and great thicknesses of sedimentary strata that sit upon the North American plate. Some thousands of metres of sediments make up much of the continental mass. As they accompanied the gorge rocks on their great journey, many other limestones, sandstones, shales, and saline deposits accumulated in surrounding regions.[9]

The Guelph dolostone came into existence in a continental area, more exactly as part of an epicontinental sea that covered much of the Canadian Shield. It was and remains far removed from plate boundaries.[10] In the great journey of the gorge rocks, the Canadian Shield remained an ever-present factor, a key to Ontario's "old land" topography mentioned earlier. Also known as *Laurentia,* the Shield is the geological foundation of the North American tectonic plate, its oldest and most stable part. It developed as resistant, crystalline rock that has survived since before the Paleozoic. In plate tectonics it is referred to as a **craton**, a massive and resistant expanse of continental crust. It is generally twice the average crustal thickness and underlain by relatively cold material that also promotes stability. As such, the gorge was formed in and occupies a stable part of the continental interior.

Dramatic in its own way perhaps, the gorge not only sits amid subdued terrain, but in an exceptionally stable part of the continental interior.[11] Eventually, and crucially, plate movements brought the gorge rocks to what would become the Great Lakes Basin. Here they were ultimately placed in the path of one of the largest ice sheets of the Quaternary Ice Age. On the ground, a seemingly simple transition marks the beginning and end of this huge episode; an **unconformity**, in which the Silurian bedrock sits beside and beneath glacial deposits from the final stages of the last Ice Age.

It is at this point that the discussion turns, finally, to the exact origins of Elora Gorge itself, and what may be called its *Third Age.*

There are many echoes of the Ice Age in today's Gorge (see Figure 3.4 for full image).

Chapter Five

Ice Age Origins of the Gorge

"But how the Grand River and the Irvine, which here form a junction, ever cut such a deep course through so many miles of limestone rock is a mystery."
— *The Lightning Express* (Elora), April 22, 1875[1]

_____

MYSTERIES

Back in the 1870s, the origins of the gorge were indeed a great mystery. In 1874, David Boyle remarked:

My own opinion is that the chasms were produced (for all I know ten million years ago) by a sudden convulsion of Nature—an earthquake—[and] the water when it began to flow naturally took the lowest level, and thus the chasms become utilized as river beds.[2]

He was not alone in attributing Elora Gorge to a cataclysm. Most early views invoked something extreme; geological upheavals or great floods. A widely accepted view blamed the biblical deluge, "Noah's Flood." Indeed, to the late nineteenth century many Earth scientists favoured the so-called Diluvial Hypothesis. *Diluvial* means, essentially, "attributed to great floods."

The diluvialists called the mounds and ridges of sediment covering much of Southern Ontario **drift**. They treated it as a legacy of the biblical flood. The sand and gravel that fills the valleys, the hills of sediment that surround the gorge, were believed to have been brought by the sea. Vast areas of similar drift are found across Europe and Siberia, as well as much of North America. They were once similarly explained by and invoked to support the diluvial theory, as well as Boyle's "catastrophism." Even as Darwin's ideas were being worked out and embraced, it seemed the very terrain of Southern Ontario favoured a biblical explanation. Once again, however, views of Earth history were about to change in an unexpected way.

Today's approaches to understanding Southern Ontario landscapes are separated from early ideas by one profound discovery in particular, the *Quaternary Ice Age*. Largely unrecognized or rejected before the mid-nineteenth century, it is almost universally accepted now. In barely a generation, diluvial theories were swept away. *Glaciation* was seen as the real origin of drift deposits. It was soon agreed that a series of huge glaciations had transformed much of the world, and all of Canada's landscapes. The amounts of meltwater attributed to the ice shifted the focus from marine to glacial floods.

This next critical phase of the gorge story—the Third— which followed from the "great journey," was ushered in by the Ice Age. It was dominated by glacial conditions. Far-reaching global climate change was decisive. Eventually, the gorge itself came to be seen as a product of these events and as much younger than Boyle imagined.

Glaciations recurred over more than two million years, on a continental scale. Ontario and the larger part of Canada's land mass were repeatedly buried under the *Laurentide Ice Sheet*.[3] At the maximum extent of glaciation, Southern Ontario was buried under glaciers up to two kilometres thick. This was the case as recently as 17,000 years ago, when the Grand River basin lay beneath the eastern or *Labrador Sector* of the Laurentide Ice Sheet. This last major expansion of the ice is known as the *Late Wisconsinian* glaciation. Its subsequent retreat accompanied special developments which were decisive for the origins of the gorge.

Improbable Terrain
Some elements of the Ice Age history of Elora Gorge are deceptive, and they reveal further ways in which the landscape can seem out of place. For the most part, surface features of Southern Ontario are made up of unconsolidated sediment. The rock canyons of the upper Grand are surrounded by hills of unconsolidated sediments, by flood plains, terraces, and alluvial streams. These are mainly glacial material left by the ice and meltwaters. If they interest you, within a few kilometres of Elora you can find a near-complete spectrum of glacial deposits. They provide textbook examples of *glacial tills*, **moraines**, *eskers, kame deltas, drumlin fields, kettle holes,* and other legacies of glaciation.[4] Professional interest has been driven mainly by the discoveries of huge aggregate resources, notably sand and gravel, and their economic importance for urban-industrial developments.

In its later, waning stages, the Laurentide Ice Sheet divided into separate lobes. These interacted in and near the Grand River basin. The more important ones are identified as the *Ontario, Simcoe, Georgian Bay,* and *Lake Huron lobes* (fig. 5.1). The landscapes left behind record the changing positions and geometry of each lobe, along with the debris loads they carried and deposited. The lobes also had some measure of independence. There were individual re-advances within

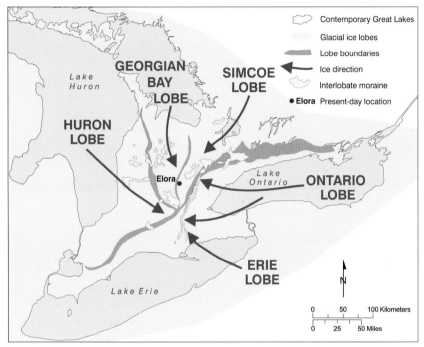

Figure 5.1   Location of late glacial lobes of the Laurentide Ice Sheet showing their relations to the terrain around Elora and beyond. Lobe boundaries record strong interactions and areas where substantial interlobate moraines were built (after Chapman and Putnam, 1984, 29, Fig. 11a). © King's Printer for Ontario, 1984. Adapted and reproduced with permission.

the general pattern of retreat. Where they encountered each other, the deposits and landforms record complex **interlobate environments**.

Interlobate Conditions and the "Ontario Island"

The development of the gorge depended upon where meltwater streams flowed and how they cut channels in rock. Of singular importance were **glacial spillways**. These developed in, or emerged from, southward directed interlobate areas. Parts of the spillways were cut in the ice, across or downstream of the ice lobes. It has been suggested that some formed **tunnel valleys**. These were most likely to initiate channel erosion superimposed into underlying bedrock. Spillways directed drainage and could further concentrate the highest flood flows. In locations where incision was rapid and in bedrock, spillways literally entrenched the early gorge.

Initially, meltwater was forced southwestward along spillways in a relatively narrow tract between the Huron and Ontario lobes. For a time, meltwater paths seem to have

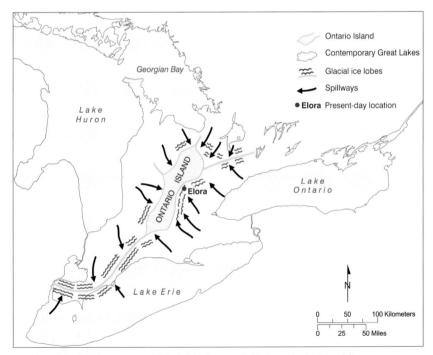

Figure 5.2   **Initial emergence of Ontario Island surrounded by the major ice lobes (after Chapman and Putnam, 1984, 30, Fig. 11b). © King's Printer for Ontario, 1984. Adapted and reproduced with permission.**

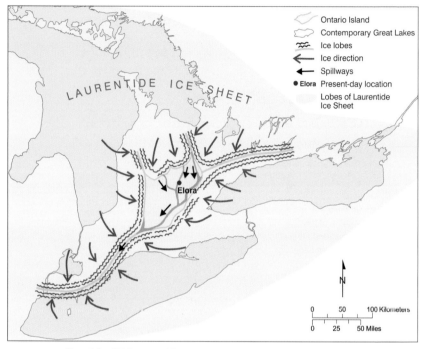

Figure 5.3   **Middle stage of Ontario Island, when its northern part was still over-ridden by the Georgian Bay lobe. Note the major spillways identified by Chapman and Putnam, especially those directed toward the Elora area (after Chapman and Putnam, 1984, 30, Fig. 11c). © King's Printer for Ontario, 1984. Adapted and reproduced with permission.**

followed the line of interference between these lobes. Outbursts may also have come from the Simcoe and Georgian Bay lobes to the northeast. L.J. Chapman and D.F. Putnam show the Grand valley above Fergus, and the Irvine north of Salem, influenced by positions and interactions of the interlobate environments.[5]

In the late glacial period, as the ice lobes waned, they left some land areas free of ice, some even for centuries before surrounding regions lost their ice. Thus, after the ice was largely gone from the upper Grand, Saugeen, and Boyne watersheds, the Ontario and Huron Ice Lobes continued to cover ground well south of future Waterloo town site and Niagara Falls.

Between 14,000 and 11,000 years ago, the *Ontario Island* was a singular development (figs. 5.2 and 5.3). This compact, glacier-free area expanded northward as the Georgian Bay and Simcoe lobes receded in that direction (fig. 5.4). The Huron Lobe defined and constrained the western flank of the upper Grand. The "Island" opened up first to the west of Elora, between the Ontario and Huron Lobes. It eventually included the height of land of the upper Grand. The gorge is located

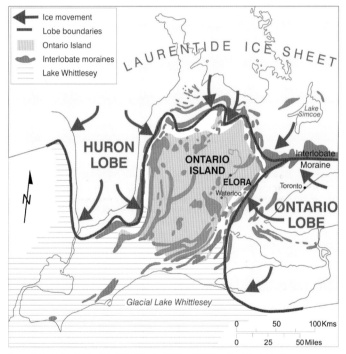

Figure 5.4   **Late stage of Ontario Island, with the Grand valley ice-free, and summarizing related interlobate moraines, the main ice lobes, and Glacial Lake Whittlesey (after Chapman and Putnam, 1984, 32, Fig. 11g). © King's Printer for Ontario, 1984. Adapted and reproduced with permission.**

in or near where spillways drained across Ontario Island.[6]
In fact, Elora Gorge appears to be, especially, an outcome of
Ontario Island and related interlobate conditions.

Glacial Great Lakes
The growth of Ontario Island as ice-free and relatively dry land
is only one part of the late glacial story. As the Island grew so
did glacial lakes (see fig. 5.4). Between about 14,000 and 8,500
years ago, a series of lakes drowned large parts of the terrain.
The more prominent lakes included Whittlesey, Maumee,
Algonquin, Iroquois, and Stanley. It was roughly around this
time when, as suggested here, Elora Gorge originated.

While the gorge seems to have been carved out on ground
beyond and above the lakes, the latter are reminders of critical
developments at that time. Not least were conditions gener-
ating the abundance of meltwater, impounded under waning glacial conditions and in unstable ponds. Some, possibly
many, channels were confined in unstable conduits and tunnels in the ice, possibly tunnel valleys. There have been some
fierce debates, still unresolved, about the full extent and influence of late glacial floods around the time of the gorge's origin. Thanks to the higher relief of the interlobate areas, stream
flow gained additional energy and higher velocities.

For the gorge to develop required an initial carving of
channels in bedrock. The important cuts had to be deep
enough, and sufficiently well connected, to remain part of the
regional drainage after the ice left. The canyons of the main
gorge had to capture the upper Grand drainage, and ensure it
went on receiving most of the stream flow down to the present. Of course, the legacy of the ice lobes continued through
the erosion of exposed bedrock and channels.

Interlobate conditions and many large lakes increased the
probability of catastrophic floods. In the wider region, studies have shown how breaching of ice or moraine dams, and

outbursts of subglacial ponds, resulted in **megaflood** events.[7] An added factor was, perhaps, enhanced or more variable snowfall due to the "lake effect" on and around Ontario Island. It is another way higher flows and floods could have been generated.

Meltwater in flood events was likely the foremost factor initiating gorge-cutting. The scale, timing, and directions of flow depended as much, perhaps more, on interlobate ice dynamics than upon climate. Changeable ice behaviour ponded or liberated meltwater in outburst floods. Flows were concentrated along glacial spillways. Hence, these late glacial conditions provided the forces and tools to excavate bedrock canyons.

Superimposed Drainage

It is important to try and imagine the landscapes of the upper Grand in the late glacial period. The influence of the ice was not merely through glacier movement and meltwater. As long as they existed, the ice lobes acted as a second, higher landscape standing over, above, and beyond the ice-buried land. Active ice supported a higher and more uneven terrain where the ice sheet or its remnants persisted. Lobes and cliffs of ice rose tens to hundreds of metres above the future land surface (see fig. 5.5 for a contemporary example). Had you stood where the land was first exposed—say, on the Ontario Island around Elora— you would be well below distant cliffs and pinnacles of ice. In places you would have had to tilt your head back to survey ice walls much higher than the gorge walls now are.

It seems the gorge, or major parts of it, are variants of so-called *epigenetic, discordant,* or **superimposed drainage**. These are streams that have been let down into bedrock from an overlying surface or land cover, to which the stream patterns were formerly adapted. They can maintain the same courses and dimensions although they now flow in a different, buried bedrock substrate, or under-mass. In the gorge's case,

Figure 5.5   Author's 2009 photo from the contemporary rim of North Patagonia Ice Cap, Chile. This indicates the kind of ice age conditions that applied in the Grand River when the gorge formed.

the latter was the Guelph dolostone. The streams being super-imposed on it were initially adapted to the glacial terrain and conditions of the interlobate environment, and places where spillways developed.

Once the gorges were incised below the surrounding ter-rain, they took control, being able to entrain and flush away most of the sediment brought from above. In the surrounding land, glacial sediments continued to dominate the landforms. Meanwhile, the rock bed gorges began to impose a separate landscape story, that of rivers in rock and a unique suite of rock wall landforms.

After the Ice; the Primordial Forest

For perhaps a couple of millennia after the ice was gone, conditions are likely to have resembled the Arctic tundra of today: a surface largely devoid of plants and lacking fertile soil. It was a landscape ruled by frost, frozen ground, and freeze-thaw cycles. This **periglacial regime** was subject to bitter, drying winds off the retreating ice sheet. In time, however, the habitat improved. The many lakes made for milder conditions. Moving northward, denser and more diverse plant life colonized the land.

Forest conditions seem to have prevailed in the Grand River basin from almost 10,000 years ago, dominated by fir and spruce. By around 5,000 years ago the natural vegetation was much like that encountered by the first European settlers. For the longest part of its existence Elora Gorge lay under the forest canopy, as did most of its surroundings.[8] I see this as the Fourth significant age of gorge history.

The forests modified the hydrology, increasing water storage across the land and in the soils that formed. The rivers would have been affected by **woody debris**, perhaps by frequent log jams in the gorge.[9] Most important, surely, was the work of beavers. Their ponds and dams tended to reduce stream velocities and flood heights. The extent and numbers of wetlands would have increased. Erosion rates must have been reduced in the gorge, coming to depend more upon rare storms and exceptional winter snowfall or freeze-up. Chemical exchanges in the limestone canyons and caves of the Grand River persisted and some of these features enlarged, but in general the pace of erosion diminished in the forest tracts.

PRESENT-DAY LANDFORMS AND PROCESSES

The gorge may be uniquely a legacy of the Ice Age, or its final phases. It has, however, continued to change and evolve since then, and remains an active landscape. Containing the steepest parts of the Grand, it involves the most extreme and vigorous erosion found in the region.

To address these matters, the focus of the final chapters returns to the living gorge and present-day processes. They may not seem as compelling as the ancient barrier reef, the so-called great journey, or the Ice Age. The focus shifts to conditions and developments happening now; landform changes noticeable in existing lifetimes.

Attention turns to those aspects of landform science most clearly involving the gorge. A variety of processes affect today's gorge landscapes. Weathering accompanies wetting and drying of the surface, and freeze-thaw days. There are various impacts of rain or snow, of winter needle ice in both soil and near-surface rock. Dust is moved around in windy weather; trees may be blown over. Tree roots and animal paths can disturb the surface or create lines of weakness. However, three sets of processes have special importance.

First there is *river work*: erosion, transport, and deposition of earth materials by the Grand and other streams, addressed in chapter 6. A second set of critical processes involves the forces that create and maintain the rock slopes, or can pull them down. Known as *slope processes* or **mass wasting**, they include falls and slides of bedrock when material breaks loose from cliffs. Here enter the conditions that support steep cliffs and resist gravity: expressions of the strength of dolostone outcrops in the gorge. These are the subject of chapter 7. Third, certain chemical reactions involving the limestone play special roles in gorge landforms: a slow but constant dissolution of the dolostone by natural waters, and deposition or precipitation of *calcium carbonate*. These processes give rise to landscape features called karst, notably cave systems, and are the subject of chapter 8.

Finally, the role of people in the landscape will be addressed. Questions of how they affect the riverscape confront special challenges, not least for the study of natural history in the "Anthropocene," a geological epoch increasingly dominated by human activity.

Figure 6.1 At a low cataract in dolostone along Irvine Creek below Salem, the channel has been extended upstream by quarrying of the bedrock, mainly exploiting weaknesses in horizontal slabs. These are lifted from the floor by more extreme flows and battering by lumps of winter ice.

Chapter Six
Rivers in Rock

Attention now turns specifically to the activity of rivers, or *fluvial* processes, mainly present-day ones. In many ways this is the pivotal chapter. Echoing the book's subtitle, it focuses on the predominance of streams flowing in, over, and beside outcrops of dolostone. This tends to emphasize erosional activity, particularly of bedrock, or **primary erosion**.[1]

RIVER WORK

As the excursions in chapter 2 revealed, Elora Gorge is distinguished by the extent of bedrock channels and rock walls. The gorge rivers exist in constant contact and interaction with the limestone (fig. 6.1). As is typical of bedrock rivers, they tend to be steeper and deeper than alluvial streams. The steeper parts of the Grand River are found in the gorge. Stream gradients tend to exceed those of the headwaters to the north.

Streams in rock have relatively fixed, erosion-resistant channel beds and boundaries. When river flow increases in the gorge, whether due to rainfall or snowmelt, there is little or no space for the stream to spread laterally. Thus, along these more confined canyons, flood waters can achieve exceptional depths, flow velocities, and turbulence, and can change quickly.

Here, observed river reaches change character frequently and suddenly, between rough and smooth sections, shallow and deep water. The bedrock channels can change direction sharply. They are disrupted by steps, rapids, sudden drops, and occasional cataracts. White water sections are common. Boulders in the stream are soaked by spray and turbulent water. As flow increases so do eddies, standing and breaking waves. In floods, these can extend the river's work onto adjacent rock walls.

It must again be noted how these conditions differ from most of the surrounding valleys and parts of the Grand River (fig. 6.2). Consider Chapman and Putnam's remarks about the Grand just below the gorge: "The alluvial plains in which the river meanders are several miles wide at the maximum. It

Figure 6.2   **View of the Grand River in Wilson Flats, just downstream of Elora Gorge.**

would be hard to find the equal of the volumes of gravel and sand in this section of the Grand valley."[2]

Moreover, if you explore the valley downstream of the gorge, between West Montrose and Kitchener, you will find it differs radically from the gorge above. Where it is not fully urbanized, one observes extensive flood plains and depositional river terraces. There are river flats and a well-developed series of river *meanders* and *oxbow lakes*, typical features of *lowland*, alluvial rivers and, indeed, of most of Southern Ontario. These features are the most widely addressed in studies of fluvial geomorphology.[3] Such landforms, including significant flood plains and alluvial beds, are rare or absent in the gorge sections of the Grand.

FLUVIAL EROSION IN A BEDROCK GORGE

The changeable world of the riverbank is a good place to begin identifying the scope of fluvial action. It is evident from previous chapters that bedrock composition and strength are critical. The bedrock surface is a crucial interface between geological and landscape functions. Bedrock channels define and constrain the space in which primary erosion can occur; notably by water coming from the river basin upstream. Bedrock constrains the lowest levels to which erosion can usually penetrate. It commonly defines what is known as **base level**: the limiting depth of the channel and elevations of stream incision.

It is important to note, however, that in karst terrain, the true base level may be displaced to deeper levels by underground drainage. Note too, how the past and existing channels of the gorge and rock terraces above, represent earlier controlling base levels of incision. The highest may even represent incisions along spillways that originated in the late glacial period.

In the gorge, mechanical erosion is the most effective agent attacking stream boundaries. Solution of dolostone at the rock-soil or rock-water interface is probably second. Bedrock channel floors and banks submerged in the flow are eroded by sediment and *clasts* (rock fragments) swept along in the river's current. This causes *abrasion*. Mechanical breakage and scour are generally in proportion to the numbers, mass, angularity, and hardness of particles transported by the stream. Impact forces and scour tend to increase with stream velocity and clast size. In the gorge, a predominance of coarser materials gives greatest effect to pounding, scraping, and chipping by stone, especially if clasts are angular, which most tend to be. The more angular the particles, the greater the abrasion effects. Larger stones may deliver crushing blows to bed and banks. Localized pitting occurs as rock chips are removed.

All other things being equal, the quantity of particles moving through, and their velocity, determine erosion rates. At a certain point, however, increasing concentration of particles tends to reduce or smother impacts and reduce scour. This may apply to the highest sediment loads moving through the gorge.

The landforms and rates of erosion also reflect differences in dolostone's susceptibility to erosion—say, between more finely-divided lagoonal strata and massive, reef-built limestone; or between stratified, slab-shaped, or rounded units. Moreover, among materials of equal hardness, rock that is heavily fractured or weathered breaks up faster than more massive, uniform material.

Low and even normal stream flows appear to have little or minor effects on the gorge dolostone. Conversely, after a flood there are always fresh-looking scars on the rock floors and submerged parts of channel floors. They resemble an incomplete jigsaw puzzle. Pieces of the stone have been lifted off, levered, or prised loose and carried away.[4]

In highest flows, extreme pressure fluctuations are generated at the bed. There can also be some especially aggressive processes such as *wedging, plucking,* and *quarrying* (fig. 6.3).

Figure 6.3   **Low cataract in dolostone along the Irvine Creek. It is being progressively extended upstream through preferential removal by quarrying and plucking of blocks where exposures exploit weaknesses in the bedrock.**

These can be sufficient to loosen and detach dolostone fragments, even large blocks. **Cavitation,** a form of rock destruction, may occur in the most extreme flows. This results from local buildup of stresses sufficient to produce vapour-filled cavities ("bubbles") in the flow. These can implode and generate local forces sufficient to eat into rock. The process is notorious for cutting into and destroying ship propellers or dam abutments, although there are doubts about the efficacy of this mechanism in nature.[5] In the gorge, if it occurs at all, it is only likely in the very highest flows.

DISTINCTIVE LANDFORMS OF THE BEDROCK GORGE

Potholes

Potholes are among the most characteristic streamlined forms, and clear evidence of the river's ability to erode bedrock. They are created mainly by abrasion, at places where it is concentrated. They are widely observed in the gorge (see fig. 6.4). Examples vary from small, fist-sized hollows to some large enough to accommodate several people standing. Smooth, bowl-shaped features, potholes are usually deeper than wide. Whereas most stones found in the gorge are angular, derived from local rockfalls and rockslides, those trapped inside potholes for extended periods become well-rounded. Potholes can also be due, in part or wholly, to dissolution of the limestone by streamflow whirling through the same paths.

Figure 6.4   **Close-up of plunge pools, rock steps, and strath terraces shaped by the river where the Grand rushes through fossil-rich reefs of the Middle Cataract or Little Falls** *(Site 14).*

Figure 6.5   **Diverse streamlined forms scoured in the bed of Irvine Creek upstream of David Street Bridge *(Site 5).* Here, as well as the irregular bedrock floor, angular boulders from rockfalls coming from the walls upstream play a part.**

Find potholes on Excursion 2, *Sites 14* and *15.*

Clusters of potholes are typically associated with cataracts and sometimes with plunge pools or elevated bedrock areas immediately above or below them. You can encounter occasional ancient potholes well above the present streams, even at or beyond the gorge rim. They also tend to be found together with a variety of other scour forms such as streamlined grooves, flutes, troughs, and scoop-like features (fig. 6.5).

Pothole development also varies according to bedrock properties. The classic bowl-like shapes occur in more homogeneous dolostone, usually where surfaces and flanks of massive reef units are exposed to fast-flowing water. Elongated scour forms may be oriented in a variety of directions: parallel to the channel and stream flow, at right angles or transverse, as influenced by local irregularities in the bed.

Find scoured stream bed forms on Excursion 1, *Site 5*; and Excursion 2, *Site 15*.

Fluvial *Notch* and *Visor* Forms

Where rivers are forced against the gorge walls, there can be marked fluvial erosion affecting cliff geometry. A range of stream-cut cavities can develop, with hanging rock above rock shelters, or solution caverns within the walls. Among the more striking are well-developed undercuts called *notches*. They can reach a metre or two in height and are box- or C-shaped. Some are deep enough to shelter a small crowd of people. A

Figure 6.6   **Multiple merged notch and visor features undercutting the right bank of Irvine Creek, seen from Lover's Leap** *(Site 2)*.

few extend tens of metres along the cliff base. The upper part, where there is a prominent overhang by the cliff above, is called a *visor*. Above this there may be an extensive free face continuing up to the crest of the cliff or canyon. In many cases there is a canopy of cedar trees, which may also overhang the visor. An impressive and accessible example is where the Irvine enters the Grand on its right (east) bank, beside *Site 7* and below the arena (fig. 6.6). The notch is carved in the base of one of the most massive reef structures exposed in the gorge.[6]

Find notch and visor features on Excursion 1, near Site 7, or view them from Lover's Leap (*Site 2*).

Details of these features and their place in landform studies are complicated. Some **notch and visor** forms have developed out of pre-existing river-cut notches. They can become the focus of interaction between several processes. Some notches are partly, even mainly, solution hollows where moisture dissolved the limestone. Groundwater seeping out at the notch or from related cavities can concentrate weathering. Still others are where freeze-thaw action is concentrated. Some notches involve spring lines. Moisture reaching the surface at the notch may already be saturated with dissolved limestone. As the water evaporates, it leaves crystals and films of calcite that eat into the rock.

Icicles, often large masses of them, are visible in the notches in winter, suggesting preferential groundwater flow. They highlight the places where abundant moisture continues to seep out. At the end of winter, fresh rockfalls may be seen scattered below the visors and within the notches. Rock fragments may be pulled away in the collapse of icicles.

In the gorge, these combined processes of chemical and salt weathering, and freeze-thaw are not only observed in notch and visor features. Nor are they confined to actively undercut river cliffs. The Stone Sidewalks described below and encountered in Excursion 3 (*Sites 22* and *23*), include developments of other, related features. However, in these the balance of control shifts from the river to rock wall forms and processes, treated in chapter 7.

Elsewhere in the world, notch and visor forms are found in diverse settings. They can develop as wave-cut notches at the base of sea or lakeshore cliffs. Similar features are well known in variable limestone (karst) settings. Certain arid and semi-arid landscapes have what are called "inland notches." They look like the coastal examples but arise from arid land conditions, also very different from the climate of the gorge.

Boulders in the Stream
While the gorge streams flow mostly in bedrock, there are channel sections with localized accumulations of unstable, usually angular, stones and boulders (fig. 6.7). In-channel erosional debris is nearly all dolostone, indicating local provenance. Such debris indicates links and interactions between fluvial and slope processes. In particular, coarse debris tends to

Figure 6.7   **Angular boulders in the Irvine, which can be moved in the highest flows.**

form and be removed or replaced in floods. The largest rockfall boulders mark extreme events that occur more rarely, but can survive and resist further movement for decades. There may be piles of gravel or rockfall that descended from the cliffs above. Finer material, even gravel and smaller boulders, tends to be removed in each year's floods or river ice events.

These clusters of in-channel boulders can seem random or chaotic. However, more careful inspection shows well-developed organization among the stones. Sometimes it seems as though someone has placed the clasts for stepping-stones, or small weirs to hold back the flow. In fact, most turn out to be products of **natural stacking**, jammed into position, stone against stone, by the force of the stream alone. They record natural assemblages related to stream flow. The stones tend to be lined up according to size and shape, usually grouped in relation to the largest boulders (fig. 6.8). The latter can act as keystones against which others are jammed. Blocks may be wedged corner against corner.

Figure 6.8   **Natural stacking of angular boulders left by a flood in Irvine Creek, with flow from right to left. The large boulder in the centre measures about 1m in its longest dimension.**

This phenomenon was the subject of one of the very few advanced, innovative studies of geomorphic processes in the gorge. In 1977, Professor Peter Martini of the University of Guelph published a study of debris movement during a powerful flood in Irvine Creek in July, 1974.[7] He showed that apparently chaotic deposits in gravel—graphically identified as *bars*, *ribs,* and *clusters* of coarse debris—turn out to display well-organized fabrics that involve overlapping or imbrication of angular clasts. Martini explained this phenomenon was the result of interaction of stones responding to sharp increases or decreases in stream flow, the process peaking as the flood event passes through.

In Canada, such forms have been recognized and investigated mainly in mountain basins of the west.[8] Nevertheless, they are found in parts of the gorge, usually where the bedrock channel steepens, narrows, or otherwise intensifies local stream flow. Declining flows toward the end of a flood event can leave new bars or ribs of boulders across the stream bed. This is another phenomenon that can single out rivers in rock, with extreme events playing a special role.

EXTREME EVENTS: FLOODS AND THE RIVERSCAPE

The gorges are primarily a result of the river's work, and of stresses developing in rock walls undercut by the stream. Powerful processes are needed to erode and carry away exposed rock; forces able to overcome the stubborn resistance of relatively hard and massive dolostone units. The young-looking channels, in many cases freshly scoured with rockfalls piled up, testify to powerful forces still at work today. Indeed, if invoking great catastrophes like Noah's flood no longer suffices, relatively rare, extreme events continue to play a significant part in the evolution of gorge landforms (fig. 6.9).

In bedrock channels, with their resistant channel and bank material, effective erosion depends upon fewer, more powerful

Figure 6.9 **The Grand in flood, with turbid waters passing under Lover's Leap** *(Site 2).*

events. Something extra is needed to attack and carve the more striking features—the quarrying out of stone steps at the cataracts, swarms of potholes in certain reaches of the gorge, and the collapse of rock walls. After the highest flows, one finds fresh surfaces where rock has been removed. New bars and snags have formed. Old gravel bars and trees may be torn up or gone. A tangle of broken trees has been thrust aside. Work elsewhere suggests that, in such rivers in rock, the greater share of erosion is accomplished in just a few events on a few days of the year. It may even occur over just a few hours during the strongest floods.

Two types of rare and extreme events are of special interest on the upper Grand today. One comprises the larger late winter or early spring ice jam floods, the other, extreme rainfall in hurricane-generated storms. In recent history, each seems to have happened only once in a century, though their frequency may have been greater in the earlier postglacial time.

It is likely to increase with global climate change. In any case, their impacts on gorge landforms can remain for centuries.

Whereas precipitation is fairly evenly distributed throughout the year, stream flows are not (recall fig. 1.5, page 13). In most years, the bulk of runoff occurs in March and April. It is associated with spring breakup and ice jam floods, sometimes aided by rapid snowmelt or spring rainstorms. By contrast, flows are usually extremely low in August. However, the most severe floods can occur at almost any time of year. Examples in recent times include:

- October 15, 1954: Hurricane Hazel, with the highest regional rainfalls and stream flows on record;
- May 17, 1974: Heavy rainstorm followed a major April snowstorm, when flood control reservoirs were already full;
- December 29, 2008: Heavy rain, with 100+ mm falling in the upper Grand Basin;
- June 23, 2017: Thunderstorms bringing the highest recorded rainfall within the upper Grand Basin.

During Hurricane Hazel, great changes occurred very quickly. Residents who experienced it in Elora say they have never heard the river roar so loud at any other time. On the Excursions described in chapter 2, you encounter boulders that were moved some way downstream by that flood—but not since. A singular impact was at a footbridge across the Irvine, below the Victoria Park steps, which once connected the town centre with the arena. Two concrete pillars supporting the bridge were swept twenty metres or so downstream, where you can still see them in the river (fig. 6.10). In all the years since "Hazel," no other flood has had the strength to move them. Information is lacking to estimate the return interval of a flood of those dimensions, but it seems likely to have exceeded the commonly cited "100-year flood," and provides a clear case of landform change that depends on rare events.

Figure 6.10   **Footbridge across the Irvine from Victoria Park to the arena, destroyed in the flood from Hurricane Hazel, 1954 (near *Site 4*). (a) The footbridge, ca. 1945 (Wellington County Museum and Archives, ph 14386), reproduced with permission. (b) The bridge pillars where they have lain since Hurricane Hazel.**

Find the remains of the footbridge pillars on Excursion 1, near *Site 4*.

It is important to recognize a series of other conditions that tend to be ignored, yet that add significantly to and complicate the landforms identified with rivers in rock. These involve the roles of vegetation, snow, and ice—notably river ice, and the phenomenon of hanging valleys.

OTHER LANDFORM-SHAPING CONDITIONS

Woody Debris

I suggested in chapter 1 that, in appearance, today's gorge is substantially a "gift" of trees. This is thanks to some decades of regrowth and spread of wooded areas along the Grand, which also reversed more than a century of deforestation.

The presence of trees promotes some forms of erosion and inhibits others. Tree roots can help anchor slopes. In colonizing and leaning out over cliffs, trees can add leverage to pull them down. Trees may be uprooted (*windthrow*) and broken off (*windsnap*), especially in heavy rain, during floods and windstorms, or due to snow loads. Along the gorge, as in forested mountain rivers, fallen and broken trees can obstruct channels with log jams. They trap and help build up coarse sediment in the stream.[9] Tree branches, burdened with snow, may be pulled down into and hinder streamflow.

Referred to as (large) woody debris, these elements of vegetation interact with the stream. They can help block or divert stream flow and accelerate or retard channel evolution. Along Irvine Creek, there are numerous irregular tangles of branches, tree trunks and roots that have collapsed down the walls (fig. 6.11). Run-out of rockslides, especially their terminal ridges and larger clasts, can interact with woody debris to create barriers that obstruct or divert stream flow.

Figure 6.11   **Woody debris, a mix of bushes and log jams along with boulder bars, impede and direct the flow in Irvine Creek.**

Today, the Grand Gorge is less prone to blockage by woody debris. Few trees are tall enough to span the main channels if they fall. However, that was almost certainly not the case before Europeans arrived. With dense forest in all directions, woody debris and forest hydrology were surely major and combined factors in developments in the gorge. So too were beavers, as remarked in chapter 5. One suspects an era of postglacial, forested conditions that involved much larger, more extensive, and longer-lived buildups behind log jams than what is now observed.

Winter, River Ice, and Spring Breakup

Winter landscapes have a classic place in Canadian awareness, and few are more memorable than those in Elora Gorge (fig. 6.12). Snowfall and river ice make substantial contributions to features, and to seasonal changes along the upper Grand. Ice buildups in the channels can aid freeze-thaw or protect the rock from erosion. However, with respect to landforms, winter conditions have received much less attention than they deserve.

Figure 6.12   Elora Gorge in winter, circa 1890, by Thomas Connon. Courtesy of Archival & Special Collections, University of Guelph, Connon Collection, XR1 MS A114302.

Plenty of research has been done on the vast tracts of lowland rivers subject to freeze-thaw and ice jam floods in the Canadian Arctic and northern Eurasia. Much is done to understand and manage river ice affecting urban streams.

You will be hard pressed, however, to find an equivalent literature on ice in rivers in rock. Yet, it is easy to see that this has a significant place in fluvial settings like Elora Gorge.

The river ice that develops and the volumes and extent of snow and ice in valley floors can be decisive for erosion, or in protection against it. Snow and ice are forced into action by sudden cold episodes and snowstorms, or they generate stream flows by thawing in warm spells. The influence of ice can extend over months of winter, largely unobserved. It may only draw attention in sudden spring breakup floods. The latter, according to my observations, are when the greatest and most concentrated erosion occurs in most years, something in which ice plays a large and varied part.

Large and small masses of ice add to rivers' erosion toolbox. Along the channels ice grows in several ways, much of it out of sight. In late fall, *shore ice* or **border ice** can begin to form along the margins of the streams. This is where the water tends to be shallow and slow moving. Freezing can happen earlier and quicker here than elsewhere. It may be aided or held in place by boulders and vegetation.

Through the winter, border ice can spread to create, or link up with, **sheet ice** which forms across the surface of the whole stream (fig. 6.13). Ice may build up slowly, or more quickly, as when **frazil ice** begins to form. The latter comprises soft, slushy masses of frozen particles that grow in supercooled surface water, particularly on clear nights with sub-zero temperatures. Some of the ice can be frozen onto stones and other obstructions. It can sink to the bottom and eventually freeze solid.

Anchor ice—ice that becomes frozen to the riverbed—is a special ingredient. At first it tends to form in shallow water, increasing in extent and thickness through cold spells. It can attach to boulders as well as the bed. Research has shown that its grip can be enhanced where there is denser, supercooled bottom water. Anchor ice may also link up with sheet ice in continuous, increasingly resistant masses and structures.

Figure 6.13 **Border and sheet ice link up to cover Irvine Creek downstream of the David Street Bridge** *(Site 3).*

Figure 6.14 **Winter ice along the walls and margins of the Grand Gorge.**

All parts of the gorge floor can become covered and anchored by ice in the dead of winter. In the longer cold spells, ice can develop labyrinthine growth patterns. There is a complex, varying mix of border, shelf, sheet, and anchor ice. Typically the frozen masses become buried under successive falls and windblown layers of snow. Broken blocks and slabs of ice may be welded together after partial or intermittent freeze-thaw events. The river ice may also become connected to frozen masses of icicles developing along the gorge walls (fig. 6.14).

In the Grand or Irvine, a shallow flow of water generally persists throughout the winter. The water makes its way along the bed, fed by relatively warm springs and seepage out of the cliffs. This means that, even in the coldest snow-free spells, ice can form and spread. As spring approaches, daytime melting and nocturnal refreezing add to the ice along the valley floor.

When the thaw arrives, especially with one main spring breakup, ice in the gorge becomes the dominant factor in erosion, and in the height, size, and incidence of floods. River ice rarely just melts in place. Being less dense than liquid water, when a thaw sets in the ice tends to be lifted off and floats with the current. Ice masses are carried along and can act like battering rams, potentially large forces of erosion. Anchor and border ice, when they pull free, can lift off pieces of rock or boulders encased in ice. Eventually, large and small slabs of ice are carried off in the stream, colliding with and blocking one another.

Ice Jam Floods and "Ice Sails"

Ice jams develop where the flow is constricted or blocked. It may be at narrows, bridge abutments, shoreline trees, rock-slide boulders, or sharp changes in channel slope. Ice jams can become stalled and form dams holding back substantial ponds. The breach or collapse of these dams can greatly increase the height and power of flood waters. Countless large ice blocks may be released and swept downstream. Ice jam floods are well-known occurrences on the Grand, usually in spring, if less frequent or large in recent years due to climate warming.

Through the 1970s, 80s, and early 90s—and reportedly earlier—there was typically just one large spring breakup in the gorge. Indeed, there used to be an annual competition for who could predict the date of break up most accurately. In the twenty-first century, however, two or more episodes of winter thaw have been observed, and several breakup floods—more events, but of smaller size. Evidently this is an effect of global climate change, accompanying a more variable and unreliable winter snowfall.

"Ice sails" refer to a mechanism that can move large boulders. This was first observed in one dolostone block, just below David Street Bridge (*Site 5*) during ice jam floods. Over several years, the block was moved in a special way. The boulder is a rough cube, three metres across. On the first occasion, it moved about ten metres along the flat rock bed during a few hours of the spring breakup. I did not identify the mechanism then. Six years later, however, and again in an ice jam flood, I saw how the same boulder was thrust along in irregular jerking motions thanks to multiple slabs of ice that were piled against the boulder's upstream flank. The boulder was pushed around twenty metres down the channel in front of ice blocks. The slabs of ice acted like a sail to increase the stress generated by the flood waters. That was in the 1980s. No further flood has had the force to move the block as of the present time.

Tributaries: Hanging and Discordant

Among other features, Irvine Creek is special as the only gorge tributary that enters the Grand at the same level; an example of what is known as an **accordant junction**. It is a commonplace of river evolution that tributaries tend to establish a common entry level with main streams. Usually, as with the Irvine Gorge, tributaries and main channel keep pace with each other in vertical incision—but this is not true of any other confluence in the gorge!

Figure 6.15 **The Cascade, by John R. Connon, showing the gorge largely denuded of vegetation, reproduced ca. 1930. Wellington County Museum and Archives, ph 17671, reproduced with permission.**

A dozen tributaries that join the Grand below Elora, and on the Irvine below Salem, are *not* concordant. Their erosion has not kept pace, or caught up with, the main stream. Their descent to it is in cataracts or, in two cases, true waterfalls. They are **discordant junctions** or hanging valleys.

A picturesque case is "The Cascade," where a right bank (Middlebrook) tributary joins the Grand in the conservation area (*Site 28*; see figs. 2.20, 2.22, and 6.15). As it approaches the gorge, the stream has carved a shallow bedrock channel less than a metre deep. This leads to and ends in an almost vertical twenty-five-metre descent into the Grand.

Another **hanging or discordant tributary** is on the opposite bank and a kilometre further upstream. It features the only true waterfall among these discordant tributaries.[10] The stream used to be fed by a small artificial lake now filled in. About a kilometre into the park, between *Sites 17* and *18*, it carved a bedrock valley and strath terraces before plunging thirty-five metres to the river.

> Find discordant junctions and associated features on Excursion 2, *Sites 17* and *18*; and Excursion 3, *Sites 20, 21*, and *28*.

Along the Irvine Gorge, between the Grand and Salem, are several more small hanging tributaries, mostly dry. A dozen such stream remnants are found along the left (east) bank, and almost as many on the right (west) flank. These suggest some disconnect between the incision of the tributaries and main streams, possibly going back to the late glacial conditions.

The hanging tributaries suggest that the gorge, and landscapes above and surrounding it, are out of phase with one another. For the moment it can be said that these tributaries have been unable to keep pace with the main river. Perhaps they occupied a basin starved of drainage due possibly to underground capture by karst streams. The main gorge has incised and enlarged itself more vigorously, and continues to do so to the Falls in Elora.

These are fluvial features typically initiated or perpetuated by changes and irregularities in stream flow, in some cases *lithology* (the nature and composition of rocks) and rock type boundaries. In many places, tectonic processes play the dominant role, but as I have already argued, they seem not to have been significant in the major gorge features. Differences between massive and finely divided dolostone may play some role. Runoff differences and extremes, including sudden emptying of glacial lakes, may have played a part once, as yet not fully erased by later changes. In the gorge they probably reflect late- and postglacial events in spillways, or subsequent stream capture. The extent, ages, and rates of retreat of such features in the gorge remain to be investigated.

AN EFFECT OF MOUNTAINS

Down in the gorge it is easy to imagine yourself in the rugged headwaters of a river in mountains, perhaps the Rocky Mountains or European Alps. The steep canyons, expanses of near-vertical rock wall, the falls and cataracts all recall the landforms of highlands. Many of the notable features of the gorge are widely known in mountain areas. And, as noted, research and researchers focused on rivers in rock have worked overwhelmingly in mountain terrain. This surely reinforces a sense of shared characteristics. However, it is also worth reflecting on the extent to which the gorge otherwise departs from this picture!

In her treatise on mountain rivers, Ellen Wohl cites eleven attributes that such rivers share. At least eight of these apply to the gorge. To paraphrase, they are:

1. Erosion-resistant and rough channel boundaries associated with bedrock and coarse clasts.
2. Highly turbulent flow with numerous along-channel transitions between sub- and super-critical flow.
3. Limited supply of sediment finer than gravel (deficit of sand, silt, and clay-sized materials).
4. Bedload movements highly variable in space and time, with higher thresholds for initiation of movement than most lowland rivers.
5. Strongly seasonal discharge regime associated with snowmelt, seasonal rainfall, or glacial melt (the latter occurring in the distant past in the case of Elora Gorge).

6. Lesser variations in channel geometry over time than for lowland rivers because only infrequent floods overcome resistance to erosion and substantial channel change.

7. Relatively narrow valley bottoms with limited development of flood plains and lateral movement by rivers.

8. Absence of wide valley bottoms and associated buffering of stream channels from hillslope processes.[11]

These all apply to Elora Gorge. However, the other three elements on Wohl's list for mountain rivers are not found in the gorge:

9. Mountain-building earth movements.

10. A river basin and surrounding terrain with large elevation differences.

11. Large vertical gradients in climatic and other surface conditions.

In the end, what separates the gorge from mountain lands is, literally, the absence of mountains. As we have already seen, earth movements have had little direct impact on the gorge rocks since their beginnings. In almost a billion years few substantial tectonic or volcanic episodes have shaped this part of Canada, something unusual in mountain terrain. Finally, in alpine environments large elevation spans lead to strong climatic and ecological gradients. Geomorphic and living conditions are vertically stacked in distinctive ecozones. These too, hardly apply to Southwestern Ontario.

What might be called "substitute" alpine conditions deserve mention: environments associated, over the past few million years, with successive glaciations. At times the Laurentide Ice Sheet, as it approached and withdrew from the region, brought not only a frozen world, but an alternative of "mountainous" terrain and climates. Ice lobes developed across the region with hundreds, even several thousands of metres in elevation. They were accompanied by conditions now seen in Antarctica or Greenland. There was great topographical relief, with regional and vertical environmental differences as dramatic as in today's high mountains. In the time identified with interlobate environments, the gorge area was shaped in decisive ways by such mountain-like features and was subject to major climate change.[12]

Figure 7.1
Vertical and hanging rock walls, and dolostone blocks fallen from them.

Chapter Seven

Rock Walls: The Strong and the Weak

Stream erosion created Elora Gorge and continues to be the main factor shaping its landforms. But the scene is also uniquely defined by limestone cliffs. Near-vertical and over-hanging dolostone walls draw the eye and record past erosion. This chapter addresses the prevalence of sheer rock faces, including the highest ones that descend directly into the rivers. Above the present channels, and in the surrounding woods, many old cliff lines and dry ravines are evident. They testify to a complicated history of cliff-forming events.

SLOPE PROCESSES AND ROCK WALLS

Relevant landform studies are identified with hillslopes and *slope processes;* land surfaces between and beyond rivers or shorelines. Hillslopes may be gentle or steep, rounded or jag-ged. Worldwide, they vary from vast subdued and near-level plains, some affected by subsidence, to steep-lands like the gorge, and onto the mountain walls of the Rockies. The steep-ness and areal extent of slopes critically affect the forces operat-ing in the landscape, mainly gravitational. Developments can also depend upon slope orientation, in relation to both solar angle and prevailing winds. The varieties and adaptive success of protective plant covers affect slope stability.

Processes that erode slopes are called *mass movements*. They include the ways in which earth materials slide, flow, or fall in response to gravity. The larger part of mass movement studies concerns deposits of soils and superficial deposits, their mobilization and movement dynamics. These have been the foremost interest of research in most of Southern Ontario. Again however, in the gorge, bedrock slopes assume special importance: in particular, erosional activity and landform evolution in places where bedrock is exposed.

Two sets of bedrock slope areas are of special interest at the sites encountered here. First are the near-vertical cliffs, many with their feet in the rivers. These are also singled out for the notable extent and varieties of hanging rock (fig. 7.1).

A second set of slopes comprise the terrain between the walls and present gorge floor, and beyond the rim of the gorge. These are generally wooded and somewhat less rugged areas, where footpaths pass through the terrain (fig. 7.2). Fallen leaves may cover the floor. Nevertheless, bedrock is at or close to the surface. There are many strath terraces and low cliffs marking bedrock collapses, old river levels, and abandoned channels. In places rockfall boulders lie at the base of cliffs, some of them temporarily lodged against trees. There is very little fine sediment.

An important landform reality of the rock walls to be aware of is their true surface area. It affects their functions as well as appearance. Measured horizontally or represented on a map, the cliffs take up very little space. They barely show up at all on aerial photographs or satellite images. However, their actual surface area is much more than it appears. A vertical wall that is, say, five metres high, appears to take up no more space than one of the same width that is thirty metres high. The latter cliff face is, however, six times greater in surface area.

The true extent of hanging rock is even more significant and challenging to determine. Being steeper than ninety

Figure 7.2   **Typical bedrock outcrops and debris-covered and treed terrain between pathways through the gorge.**

degrees, the horizontal extent inside the cliff line is completely obscured on a normal topographic map. I estimate the lateral extent of near-vertical cliffs present in Elora Gorge equals and possibly exceeds twenty kilometres. In the main gorges of the Grand and Irvine, the total area of overhung cliff face includes overhangs rising vertically several metres, and up to thirty metres in the highest parts. Where there are separate faces along both rims of the gorge, the exposed rock face area can be double or more than that of the gorge floor. In addition,

Figure 7.3 (a) and (b)
Examples of hanging rock and rock shelters typical of the smaller cliffs in the gorge.

one needs to be aware of the countless low cliff sections and minor overhangs in bedrock. Even in these cases, vertical or overhanging cliffs are the rule (figs. 7.3a and 7.3b).

CLIFF FORMS

Beyond area and steepness, the varieties of three-dimensional hillslope forms are noteworthy; the sculptural profiles of sheer and undercut cliffs. About half the sites and sets of overhangs have rectangular profiles. Slope crowns or crestlines tend to be defined by horizontal slabs and undercut ledges. They resemble architectural forms like the projecting eaves or cornices of buildings. I call them **"cornice rocks."**

Others have rounded profiles. From a distance they resemble ceramic forms like urns, jars, bowls, and vases. These I call **"potter's rocks,"** not least to acknowledge the many fine potters who have worked in Elora, and the ceramic forms displayed in their shop windows.

It is no accident that gorge forms recall the well-known Flowerpot Island on the Bruce Peninsula, overlooking Georgian Bay. It is 100 kilometres to the north of Elora and also has hanging rock that reminds people of ceramic pots. More to the point, like Elora Gorge, it is carved in dolostone and, indeed, is an outcrop of the Guelph Formation. In general, the reef features, marine growth forms, and colonies described for the Guelph dolostone prefigure both potter's and cornice rocks. Adding to the variety, some high walls have both angular and rounded forms in the same profile (fig. 7.4).

Recognizing and naming landform features helps one to appreciate and compare them. The terms "cornice rocks" and "potter's rocks" prepare the eye for unfamiliar cliff shapes and their variations. Their geometry and extent are integral to the unique appearance of the gorge. The vertical and overhanging forms are direct indications of rock that has considerable internal strength. Many overhangs are undercut by the river

Figure 7.4   Two types of vertical cliff profile observed in the gorge, "cornice rocks" and "potter's rocks."

Figure 7.5   "Potter's rocks" fashioned of reef masses undercut by the Grand, between *Sites 13 and 17.*

(fig. 7.5). Some are visible where spring lines occur at the base of a cliff, or solution weathering and freeze-thaw eat away at the base. They include varieties of the notch and visor forms introduced in chapter 6.

The cliffs record incision and removal of rock by the river. However, once present the walls, in turn, affect the river's behaviour. These relationships can be identified with three cliff conditions: *active, inactive,* or *relict. Active cliffs* are those where the present stream reaches and washes the base, and/or where fracturing and movement are evident in the cliff profiles. There may also be a concentrated solution of the limestone, rock abrasion, and quarrying, as described in chapter 6. In other places, the river is held back from the cliff by boulders, gravel bars, or treed islets. The relation is *inactive* or episodic. This marks periods when the walls persist with little or no erosion; at least until and unless a flood arrives that reaches the rock face. *Relict* features are old cliffs now set back from and above the reach of the river. They are still subject to erosion by slope processes but exist largely independent of the rivers.

The cliff itself is a natural form of strong architecture. It stands in walls comparable to, and that helped inspire, those of the great fortresses and cathedrals. The steepness and height of the walls reveal their resistance to the gravitational forces that would pull them down. They share the elements of structural strength that must be obeyed in defensive ramparts and tall buildings. Only strong rock will stand in near-vertical cliffs, the more so if it is hanging rock.

"STONE SIDEWALKS"

Between the Middle and Low Bridges, hidden away in woodland along the left bank of the Grand, are extensive slope features associated with old, relict cliff lines. They are the Stone Sidewalks, introduced earlier in Excursion 3. Here they are considered in a little more detail.

Figure 7.6   **A "Stone Sidewalk" (between *Sites 22* and *23*).**

There is generally a "path" that runs close beneath each cliff. On the downslope flank are piles of fallen boulders and weathered debris. These help to create a terrace at the base of the cliff. Above, the paths are sheltered by a projecting roof of overhanging bedrock; slabs or bluffs of dolostone. The roof lines are typically an extended series of cornice or potter's rocks. They shield the "sidewalks" from the midday sun, from rain on wet days, and from debris falling from above. Chemical weathering of the dolostone is involved too. Some of the angular rockfall debris is held together by cement precipitated from solute-rich moisture seeping out of the limestone. A berm of this material forms the path that winds its way below the cliff face.

Some paths are so smooth and walkable they almost seem artificial. Or one can imagine them as secret paths through the woods of the gorge. And perhaps they did serve First Nation hunters or traders, coming and going between the Niagara and Bruce Peninsula. Yet, whatever use humans may have made of them, the Stone Sidewalks are natural features. They derive from fortuitous interactions between rock walls, former river levels, and centuries of freeze-thaw and salt weathering (See figs. 2.21 and 7.6 for examples of "Stone Sidewalks").

The sheltered undercuts and cliffs are of modest height, generally rock terraces five to fifteen metres high. Most represent former river levels. Occasionally a large rockfall or clump of trees may obstruct the path. You must circle around them, briefly stepping out of the stone canopy's shade. In some places there are as many as three of these cliff lines between the present river and the flatter land beyond the gorge rim.

Find the Stone Sidewalks on Excursion 3, *Sites 22* and *23*.

ROCKFALLS AND ROCKSLIDES

Sheer walls and hanging rock tell of strong material, resistant to gravity. However, debris piled up at the base of cliffs tells another story; one of failed strength and processes able to destabilize and transport collapsed material. A common source of instability in a bedrock cliff is the presence of fractures, *bedding planes*, and other **partings**: lines of weakness developing in the rock itself. Their significance affects the size and geometry of failed rock units and depends somewhat on how partings dip relative to the slope. Partings also decide where moisture can

Strong

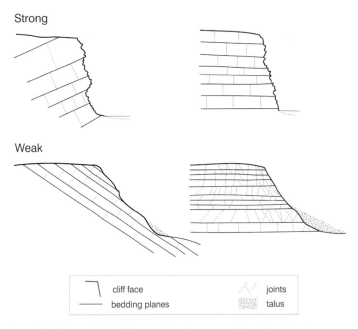

Weak

| cliff face | joints |
| bedding planes | talus |

Figure 7.7   **Structural and cliff faces, schematic sources of strength and weakness.**

penetrate more readily into the rock. These depend, in turn, on the presence of weathered and porous stone, and how it is affected by snow loads, freeze-thaw cycles, or tree roots.

Partings in the Guelph dolostone vary widely in concentration and origin. They involve all the elements in the diagram in fig. 7.7. Specific forms and relations go back mainly to the influence of marine organisms, sedimentation, and patterns in the barrier reef system. The most massive reef units display unusual strength. Thanks to the limited earth movements in the area, folding and faulting are minimal. Stratified rock units that did form remain in the near-horizontal positions of the original deposits, which also promotes strength. Even so, heaps of stones and dolostone boulders are encountered throughout Elora Gorge.

What distinguishes these mass movements is, once more, how they are derived from, and composed of, the local dolostone. Nearly all boulders and other rock fragments are angular. **Sharpstone** fragments are common, as are broken edges and signs of fracture or breakage. This is indicative of local origin in dolostone outcrops and bouts of aggressive erosion. Looking at the larger blocks, you will see how most have elongated shapes; those of slabs, rods, or cubes. The shapes are determined by release and breakup along pre-existing joints, *bedding planes,* or pressure-release fractures.

The contrast with the surrounding countryside could hardly be greater. There, if they are exposed at all, stones have diverse composition, mixing various rock types in the terrain over which the ice sheet travelled. Clasts are also mostly rounded and compact, indicators of wear in transport by water or ice. A few glacial **erratics** have been left in the gorge, far-travelled and well-rounded boulders, but they are infrequent compared to local dolostone clasts.

The verticality of the gorge walls and strength of the rock favour a narrow range of mass movements. Rockfalls and rockslides prevail. As a whole, mass movements can be divided into those that creep, flow, slide, collapse, subside or fall.[1] Generally, rates of travel tend to increase from creep to falls. Although the gorge cliffs seem relatively clean, there are countless small rockfalls in every year. The fallen stones range from large boulders to small fragments. These, and their impacts, are found below sheer and overhanging walls. The more conspicuous movements are slab- and block-falls. The larger boulders are a metre or more in diameter, and their geometry usually reflects lines of weakness in the dolostone. The prevalence of rockfalls and rockslides reflects the dominance of relatively fast-moving and short-lived events in Elora

Gorge—another way in which the processes here differ from the surrounding, gentler, soil-covered terrain.

> Find prominent rockfalls and rockslides on Excursion 1, between *Sites 4* and *5* and at *Site 7.*

CREVICE CAVES AND THE SELF-PROPAGATING GORGE

Gorge cutting itself alters and redirects the forces at work in the surrounding bedrock. River work and slope processes are linked in the co-evolution of the gorge. Consider, first, the situation where there is bedrock *in situ*, meaning where substantial dolostone masses are present. They load and compress the rock beneath; support or buttress surrounding rock masses. The weight of limestone, its thousands of tons, create forces bearing down and laterally. However, suppose it is removed, whether by river or rock slope erosion. The rock that remains in place above or on either side of a given rock outcrop will be subject to stresses within. Bedrock removal means the cliffs are subjected to tension or pull-apart stresses. The channel floor, relieved of the overlying rock burden, tends to bulge upward. The stresses weaken and may fracture the floor. The walls tend to creep or bulge outward.

Such disturbances result in an elastic response of exposed rock and, possibly, brittle fracture. Horizontal and vertical fractures may open up, aiding in excavation of the gorge. Most important are distinctive failure patterns observed in so-called crevice caves, introduced in Excursion 1 along Irvine Creek. These stresses can generate critically important vertical stress-release fractures, more or less parallel to the rock wall. Crevice caves are best-known from observations along the Niagara Escarpment. They are rarely mentioned at Elora Gorge, yet they are widely present and key factors in gorge development.

Figure 7.8   A crevice cave and major rockslide waiting to happen near *Site 4*. At the left side, the block has detached sub-parallel to the gorge wall, where a vertical tension fracture developed. The block has rotated outward at the base and is sliding down a steep slope toward Irvine Creek.

The classic crevice cave involves a planar fracture, behind but roughly parallel to the cliff face, opening up through gravity sliding (fig. 7.8). In some cases, concave fractures arise and may produce a scoop-shaped rotational rockslide. There is usually an overhanging face and, where fractures penetrate inside the cliff, they open out laterally inside. The cave may be roofed over or penetrated by vertical fractures parallel to the gorge wall.

> Find a crevice cave on Excursion 1, between *Sites 4* and 5. Refer to figs. 2.8 and 2.9, pp. 38–39.

Crevice caves have been attributed to **permafrost** and *periglacial* conditions at the end of the Ice Age. However, while that may once have been a factor, it is not a necessary one. Crevice caves in the gorge developed long after periglacial conditions ceased, and continue to develop today. The same is observed on the Niagara Escarpment. Also, crevice caves may be combined with *solution caves*, but need to be distinguished from them. There are clear differences between forms developed by chemical erosion and those caused by mechanical failure. But they may overlap and, together, they remain integral parts of present-day processes.

Stress-release fractures arising in the gorge walls and floor are critical ways in which river work and slope processes are linked. Rockfalls from such fractures help expand the gorge in ways also reflecting and affected by the shape of the canyon. Along with the crevice caves they record the fact that the gorge is a self-propagating feature. Slope and fluvial processes continue to enlarge the canyon. Meanwhile, where the blocks or slabs of rockslide materials reach the valley floor, they tend to direct river flow toward the opposing wall and protect the source slope.

EXTREME SLOPE EVENTS AND UPHEAVAL IN THE IRVINE

As with erosion of the riverbed by floods, the greatest mass wasting events tend to involve rare extremes. Over some thirty years, I have recorded eleven rockfalls in the gorge exceeding thirty cubic metres in volume, the largest in the 1970s below

Figure 7.9   **Rockslide in the Irvine Gorge, between *Sites 4* and *5*. On the gorge wall above is the detachment zone, whence the slide collapsed. Low down are blocks that rotated as they travelled down the face.**

the arena (see fig. 2.9 in chapter 2, page 38). Some remnants of prehistoric rockslides include much larger events, but sizable mass movements evidently continue to occur. Human activity in the past two centuries may well have affected slope stability, but the largest events appear to predate European settlement.

As noted on Excursion 1, many larger rockslides are found along the Lower Irvine Creek. Between Victoria Park steps (*Site 4*) and David Street Bridge (*Site 5*) is a series of overlapping slide deposits, each contributing multiple dolostone blocks (fig. 7.9). Together they make up an almost continuous set of slides that overlap. In all, they provide a veritable museum of rockfall and rockslide features in dolostone. After breaking away, the bedrock units split and broke up further as they descended, before coming to rest. The slabs of bedrock, descending rapidly from the wall, have rotated and split apart before spilling into the river. With careful observation you can detect how the slide masses interacted. Beneath the largest slabs you see rock that was more severely fractured and crushed by the overlying blocks. But what is their story?

Prior to the collapses, the Irvine flowed against the wall on this left bank. There are abundant indications of the river undercutting the cliff base, the exact opposite of today. Now the Irvine follows and undercuts the right bank. Another question arises around the causes of the concentration of rockfalls and slides here.

It is conceivable there were multiple slides, each occurring separately. More likely, I suggest, is that most, if not all, comprise a single episode of slope collapse. Without reliable dates it is hard to be sure of when that happened. The oldest trees that have colonized the rockslides seem to be of similar ages. The trunk diameters of the largest appear to equate them with known 200-year-old trees. That would predate the first European settlement.

Recent history suggests rockfall and rockslide events mostly follow the spring thaw or rainstorms, but tend to be relatively small, isolated events. For so many large rockfalls to occur together, as is evident along the Lower Irvine, suggests a single powerful trigger. If this were a mountainous region, the first likely choice would be an earthquake. Seismic activity is a well-documented trigger of landslides. But the gorge is distant from seismically active areas. In living memory, earthquakes have been rare and small. However, that is not the whole story.

In 1811–12, an earthquake episode that is thought to be the most powerful ever recorded east of the Rockies originated at

New Madrid, Missouri. It consisted of a series of large earthquakes and several powerful aftershocks, felt over an area ten times larger than that of the widely known 1906 San Francisco quake. The overnight shocks woke people from Detroit to the muddy town of York (later renamed Toronto) and were felt as far away as New York, Boston, and Montreal.

That said, it may seem odd that only this 400-metre stretch of the Lower Irvine Gorge was so strongly affected. Possibly, however, it had been uniquely prepared by the river's undercutting and the state of the cliffs. A special consideration is how the geometry of such cliffs, notably hanging rock, can locally intensify earthquake shaking. Referred to as *topographic enhancement*, in a few mountain areas it is invoked as a main factor in some extreme collapses and high concentrations of rockslides.[2] Perhaps it was so at Elora Gorge.

Figure 8.1
Cave in dolostone along Irvine Creek gorge, due partly to chemical and partly to mechanical weathering. Compass included in image for scale.

Chapter Eight
Karst

Another set of conditions that have shaped gorge landscapes involve chemical reactions, mainly *solution* and chemical *precipitation* of limestone. These add to and complement river work, as well as the stresses in cliffs and mass movements. They have special sensitivity to earth surface conditions and, at depth, to groundwater.

DOLOSTONE AND THE CHEMISTRY OF ELORA GORGE

As established in chapter 3, the main constituents of the dolostone are *calcium carbonate* and *magnesium carbonate*. They signal the paramount role of carbon, notably carbon dioxide gas, in limestone formation and dissolution. The dolostone may be somewhat less soluble than the calcite it replaced, but it is still among more readily dissolved rock types. The processes are largely invisible, but their landform contributions can be pervasive. Among the more obvious landscape features are caves (fig. 8.1) and **sinkholes**. Springs issue from the base of the cliffs, and countless small solution pits and holes are exposed (fig. 8.2).

Figure 8.2   Multiple solution pits and holes, or "vugs," in reef rock. Many are clearly identified with fossil shellfish and other Silurian life forms.

The Guelph dolostone has been characterized as "vuggy"; that is, marked by a wide variety of pits, clefts, and hollows. **Vug** is an old mining term brought from Cornwall, referring to any small cavity. The diverse shapes, textures, and density of vugs mostly exploit or "ghost" the lifeforms and organic deposits inherited from the barrier reef. They increase porosity and open up lines of weakness through the rock. They help to direct the flow of both surface and groundwater.

Find vugs on Excursion 2, *Site 16.*

Apart from the main rivers, there is little surface drainage in the gorge. Rain and snowmelt tend to sink into the ground. Water moves along underground passages. The water passing through and out of the dolostone contains dissolved carbonate. This explains why the local water in Elora is notoriously "hard." Residents must find ways to soften it, clean up the scales and stains from the carbonate, or obtain their drinking water from other sources.

Ground and soil moisture may become saturated with *calcium carbonate* and support a second set of chemical reactions that release and precipitate carbonate minerals. The extent and pace of dissolution, and of precipitation of mineral matter, depend upon the acidity of the water, in turn influenced by organic materials and activity as well as water volumes and flows.

KARST

In landform studies, features related to solution and precipitation in limestone country are identified as karst. This German term derives from the *Kras* scenery of Slovenia, on the northern shores of the Adriatic Sea. Typically, karst involves a prevalence of underground drainage, solution-derived conduits and caverns. There can be abundant secondary mineral deposits, including those in caves called **speleothems**. These include **dripstone** forms, of which **stalactites** and **stalagmites** are two widely mentioned types. There are also **flowstone** forms, comprising sheets of *calcium carbonate* on cave floors, suspended curtains, or films below roof areas. "Cave crystals" may form, including needle-like growths of calcite.

The Guelph dolostone involves many sites where the limestone is dense, massive, or coarsely fractured. As noted in chapter 3, it is of uniform composition. These conditions are said to favour the development of *karstic* features.[1] And the latter are present to some extent in most parts of the gorge. However, they are generally in small-scale features, and hardly compare with the classic karst landscapes' impressive caverns, underground rivers, gigantic sinkholes or **dolines**, and diverse dripstone forms.[2]

Meanwhile, throughout the life of the gorge, physical water erosion—abrasion, corrosion, and impact by sediment loads, already described in chapter 6—have competed with, added to, or limited full development of solution processes. What has developed is mainly sensitive to the architecture of the ancient coastline and barrier reef. The features most common in the gorge can be referred to as **bare karst**, encountered in bedrock outcrops with little or no vegetation or soil.[3] Bare karst occurs mainly where surface deposits were flushed away, perhaps, as in the gorge, along the original glacial spillways, or swept from unstable and steep cliffs. The alternative, **mantled karst**, is buried under superficial sediments and vegetation. Much of the limestone beyond the gorges is mantled karst, mostly covered by glacial sediments. Over time, of course, and with changing circumstances, bare karst may become covered in sediment or plant growth, and mantled karst may be buried in soil and plant life, complicating the interpretation of given forms.

Almost all the rain that falls within the gorge quickly infiltrates into underground drainage, as does most snowmelt. This is typical of karst terrain. Caverns and springs emerging from cavities, large and small, are common in the gorge. They involve many of its most notable landforms. The water flowing through comes largely from upstream tributaries of the Grand and is passed on quickly to downstream areas. It is accelerated by the steepness and greater depth of the channels in bedrock. However, stream flow from and supporting them is generally limited or intermittent, compared with tributary rivers.

As in all aspects of landscape evolution, questions of time are key. The world's more spectacular karst systems have developed over millions of years. In geological terms, most of the gorge karst is young, mainly formed during and since late glacial conditions. However, underground solution, cave, and speleothem development may have gone on at depth in the Guelph dolostone over the 400 million years since its Silurian beginnings. Some of the cavities and precipitated forms now being uncovered may go back even to the earliest times, others to inundations and cave formation during the Great Journey (see chapter 4).

THE CAVES

Caves draw the eye and create a sense of mystery. They are among the most intriguing places to visit and can help unravel the story of the Elora Gorge. About a dozen of its caves are large enough for a human to enter; there may be more at depth and inaccessible behind cliff faces. In a few cases, half a dozen people can climb inside (see fig. 9.4, page 184), but even the largest gorge caves hardly present a challenge for the dedicated *spelunker*.[4] This may help explain why local karst has attracted little interest—a pity, since research into karst elsewhere has involved some of the most impressive discoveries and notable landscapes in modern Earth science in Canada and beyond.[5]

Smaller cavities are abundant in the gorge; many are impenetrable or just glimpsed in openings on the cliff faces. They may be relatively minor, but they add to the scope and variety of cliff forms. They have played a large overall role in gorge karst. Some caves tap into old underground streams. Some seem to have been exposed with the cutting of the gorge, and others captured by it. There are smaller caverns that open out within the rock and provide protected ecosystems. Some remain unfrozen in winter and relatively warm, serving as havens for bats, spiders, and various insects. Some sit high and dry above the existing groundwater table. They may have been exposed by natural gorge deepening, partly in response to past climate change, partly in response to human water extraction.

COMPOUND CAVERNS

None of the larger gorge caves, and few of the smaller caverns, are pure karst forms. Equally significant are roof collapses or rockfalls and rockslides from the gorge walls. As noted in the previous chapter, some of the larger cases involve crevice caves, and related mass movement processes. Caverns produced purely by solution may be more important in deeper, inaccessible tunnels. Those below the water table may give passage to substantial groundwater streams.

Figure 8.3
One of the largest caves in the Irvine Gorge, issuing above the level of the present gorge floor, and influenced by crevice cave processes as well as solution.

However, the nearer the caves are to the gorge walls, the more they show effects of gravitational stresses. There are tension fractures in high, steep cliffs already mentioned in chapter 7. In places, quantities of fractured and frost-weathered rockfall debris descend from the cliff face or cave roof above. On cave floors and at their mouths there is more evidence of rockfall relating to freeze-thaw or wedging by tree roots. Cave roofs add to the presence and importance of hanging rock.

HANGING CAVES?

A notable feature of the larger caves in the gorge is how the main chambers open into the gorge a metre or more above the present river channel. They comprise a form of **hanging caves**. At the entrance they have a sudden drop, even a short cliff (fig. 8.3). This seems to reflect and crudely measure the irregular history of gorge cutting. The deepest and highest main chambers sit some metres above the present gorge floor. They may represent a period when heavy forest cover prevailed and promoted higher water tables and long-term stability—following the greatest and most rapid incision by glacial meltwaters.

As indicated in chapter 5, the forests were likely accompanied by much-reduced erosion. A hiatus in downcutting may mark a time when tree-fall and beaver dams obstructed rivers and tended to even out stream flows. There are indications of a milder climate between about 7,400 and 1,200 years ago; apparently warmer but also wetter than earlier in the twentieth century.[6] The last couple of centuries have likely marked another acceleration of erosion due to human activity in the watershed.[7] Among the results, this has changed relations between water tables, cavern drainage, and erosion of the main river channels. It has brought a new period of relatively strong incision by the river along the gorge, and enlargements of caves.

Figure 8.4   Spring line features in the gorge where aprons of precipitated carbonate or tufa occur. In this case the spring is smothered by mosses.

TUFA DEPOSITS AND "ROCK GARDENS"

Some of the more intriguing karst phenomena arise with local buildups of porous calcareous deposits called tufa, or "cold" travertine (fig. 8.4). These have formed around the exits of many of the caves and where springs issue at the base of rock walls. The most conspicuous buildups are deposited where there is abundant spring water. Where there are active year-round springs, tufa aprons up to several metres across are found. The water may emerge under some pressure, or in small waterfalls at the base of cliffs. Turbulence can promote aeration, which diffuses carbon dioxide, increases evaporation, and supports plant growth; all conditions which, in turn, promote tufa deposition.

Where springs emerge at or near the floor of the gorge, tufa deposits are also identified with communities of grasses, flowering plants, mosses, and algae. *Microbiota*, which flourish in the lime-rich spring water, are thought to be an important factor aiding this phenomenon. Plants decorate the tufa aprons and fallen fragments of rock to create natural "rock gardens" or "hanging gardens" (figs. 2.23 and 8.4).

Where sunshine can reach the gorge floor, the richest examples involve active buildups of porous tufa and greenery that grows year-round. In winter, a helmet of ice envelops the tufa apron and plants. Relatively warm spring water continues to flow beneath the icy crown. Light filters through the ice and may even be concentrated by it. These features are sensitive to water table height and underground drainage. When the flow temporarily fails or is intermittent, the tufa dries out, solidifies, and may become fossil.

Active springs and tufa gardens are much more prevalent along the right (west) wall of the Grand Gorge below the junction of the Grand and Irvine. As noted in Excursions 2 and 3, there are more, and more vigorous, springs here, with attendant icicle growth in winter. There seems to be a structural

Figure 8.5 A tufa "hanging garden" where some springs issue at the base of the rock wall, right foreground. Owing to variable drainage, the feature combines active, inactive, and fossil tufa.

component related to the westward dip of the rocks. Mainly it reflects the greater height and steepness of the west wall.

The tufa masses and "gardens" are not built solely of precipitated carbonate. In aprons spreading outward from the rock faces an important part consists of weathered and broken pieces of rock. The tufa encrusts and fuses these together, again suggesting that they are natural "rock gardens." The growth, but also the breakdown of the tufa masses can be relatively fast, depending primarily on water supply. They survive if spring water is reliable but dry out in a few waterless months or where the water table drops below the site. After a year or two of dry conditions, tufa can begin to turn to stone. A return of water flow can rejuvenate them.

The tufa masses, and the conditions they depend upon, reflect the hydrological history of the gorge, and its present status (fig. 8.5). Tufa gardens are affected by water pumping and well extraction for human uses. Examples of all the stages of tufa aprons, and transitions between them, are found along the Irvine. Around Salem and Elora, groundwater extraction, artificial drains, tree growth and removal have meant changeable water tables and spring lines. However, the information on these changes is mostly anecdotal.

> Find tufa aprons and "hanging gardens" on Excursion 3, *Sites 25* and *27.*

Tufa deposits not only respond to, but help construct, the gorge landscape. They add to carbonate features which reflect changing conditions since the end of glaciation. Heavy tufa deposition probably followed the rapid gorge-cutting of the late glacial period, long before serious human intervention in land use and drainage. Times when beaver dams and log jams limited gorge-cutting encouraged tufa deposition.

A perspective on the varieties of relict tufa features can be seen beneath David Street Bridge on the Irvine (fig. 8.6). Tufa deposits take up space where caves open into the gorge. They support rock benches, tufa, or strath terrace forms, beside the river. Here are recorded past conditions where underground streams reached the surface and tufa was deposited. There had to be springs at that stage, but also sites where weathered blocks fell from the roof, and lime-rich water dripped from the walls. Since then, the sites have dried up. The once-living tufa turned to stone. Eroded masses keep falling into the river.

In sum, karst processes and features have added much of the detail and diversity of the visible gorge. The contribution of solution and precipitation of carbonate, however, seems to come third in the overall balance of processes. River work comes first, initiating and controlling landscape forms and

Figure 8.6　**Tufa deposition and relict masses associated with a cave and eroded "rock garden" beneath David Street Bridge** *(Site 3).*

opening up the gorge. Second are gravity-driven rock wall stresses, falls, and slides.

It also bears repeating that all three sets of processes interact and are at work at any one time, if at differing rates. Each helps reinforce the sense of the gorge as a set of landforms in bedrock. Locally, they comprise exceptionally steep terrain. They have complex geometry, histories, and recurring bouts of extreme forces, especially along the main rivers in rock, and on the hanging walls.

A FAMILY OF LOST AND BURIED GORGES?

There are many and varied exposures of Silurian dolostones across Southwestern Ontario, and diverse gorges, notably along the Niagara Escarpment.[8] However, the closest similarities to Elora Gorge are in certain gorges, otherwise hidden features called **buried bedrock valleys (BBVs)**.

As the name suggests, these are former bedrock valleys filled in and covered over by later deposition. One of the earliest to be identified, the Elora Buried Valley, intersects Elora Gorge near the Falls and the Mill Inn.[9] Others are to the east, under the city of Guelph, and at Rockwood.[10] Some have been found north and west toward Mount Forest and Wingham.[11] A number are in the Grand River Basin, others in adjacent basins.

Recognition of BBVs began with work on groundwater and evidence from well records in the region. These also provide extensive data on the bedrock subsurface, and the location and extent of subterranean aquifers. Of late, scientific interest in BBVs has focused on their place and role in groundwater resources. They are at risk from competing extraction uses and groundwater contamination. And some are treated as important karst phenomena: legacies of limestone solution and underground drainage, as well as late glacial drainage.[12]

A number of these BBVs exhibit features closely resembling Elora Gorge, except for being buried. Developments suggest a similar late glacial origin among the more deeply incised spillways. As at Elora, BBVs have gorge sections that commence and stop abruptly. Others have convex profiles and reverse slopes, hanging valleys, or sites suggestive of former cascades. Their bedrock forms include irregular and sudden changes in channel widths, in depths and gradients, and in orientation of valley axes. Drill cores in some cases show that BBVs have masses of coarse angular rubble that were piled against gorge walls before they were buried. Evidently their walls were subject to rockslide activity and other mass movements, as well as *glacio-fluvial* processes.

The BBVs seem to have been filled with sediment in the late glacial or early postglacial times, and in some cases possibly drowned in a final lake phase. These suggest chronically disrupted rivers, subject to discontinuous and irregular events. The context for BBVs was also created by glaciation and unstable ice margins; the high and changing relief of the waning ice sheet described earlier. The features are even sufficient to suggest a family of streams, all formerly active rivers in rock.

Of late, more detailed information has become available thanks to a range of geotechnical methods and officially funded surveys. However, these investigations are not readily accessible to or understood by non-specialists. Various scientists have attempted to reconstruct the BBVs as part of pre-glacial or "ancestral" drainage systems, possibly arising from limestone caverns or, perhaps, organized by pre-existing structural and tectonic events.[13] In part, these arguments arise from a conviction that special or extreme conditions had to be involved, as with other work invoking glacial floods, including megafloods. This can refer especially to interlobate conditions, the Ontario Island, and glacial spillways. It does not preclude some of the BBVs being legacies of earlier but similar glacial episodes. I am inclined to accept the view that late glacial conditions favoured gorge cutting and subsequent burial by heavy glacial deposition. Cunhai Gao favours a similar picture, but suggests that some BBVs began as tunnel valleys, excavated by subglacial meltwater.[14] This has been a preferred view, and some believe it could apply at Elora Gorge, but I have seen no definitive evidence for that.

Growing knowledge about BBVs should contribute to understanding of Elora Gorge, which in turn preserves features observed when bedrock gorges remain exposed. Meanwhile, the one great difference is how the gorge landscapes, rather than being buried, are still gloriously open and visible, long after the ice left.

WINDOWS INTO THE PAST

As described so far, Elora Gorge provides glimpses of a story revealed mainly by erosion. It offers a series of windows on the long-term history of its region and the upper Grand River basin. A distinctive picture emerges that differs from most of surrounding Southern Ontario. Beyond the gorge, landscape history is recorded mainly in landforms composed of, or built by, moraines, other glacial features, and successive lake beds. The landscapes away from the gorge mainly record the legacies of alluvial streams, large lakes, windblown dust or sand dunes, weathering, and mass movements. In most other areas, landscapes are composed of mosaic-like deposits of surface materials—built up, that is, rather than carved by erosion, as in the gorge.[15] True, the Shield to the north and the Niagara Escarpment are also dominated by bedrock features, exposed and shaped by erosion—but again, including some marked differences from the gorge.

Such strands of its history, strictly its natural history, set the scene for a final age of the gorge. It reveals an ever-greater emphasis on the roles of people in the landscape; how they have become ever more assertive builders and excavators of the terrain. In the final chapter I will explore what amounts to the last of the ages of the gorge.

Ages of the Gorge

Each landscape has its own distinctive features and history. Erosion of Elora Gorge has exposed legacies of geological eras spread over almost a billion years. Also, several pre-gorge worlds developed, some long before anything identified with the existing gorge. As revealed earlier, the region was birthed with, and persists in the midst of, the North American Craton. This, and notably the Michigan Basin and Algonquin Arch, determined the main structural elements of the upper Grand. Thereafter, what may be called various "Ages of the Gorge" emerged. They make for a complex story, sign-posted by established geological eras and events. In summary:

- *The First Age: in the Silurian "Guelph Sea."* This first age involved the origins of the gorge bedrock as reef limestone. Laid down some 430 million years ago, its characteristics derive from a tropical marine environment. The composition and layout of the rocks, especially the barrier reef, have uniquely constrained the features and evolution of the gorge. Read more in chapter 3.
- *The Second Age: the "Great Journey."* In the second age of the gorge, a set of geological forces carried the reef limestone from a tropical marine environment to the heart of a continent. Ultimately, it occupied what became Ontario's snowbelt. It was a largely passive transport, emerging with developments in the Earth's crust, now attributed to plate tectonics. This age might equally be called

"The Great Silence." The journey of more than 400 million years and thousands of kilometres is not directly evident in the gorge landscape. The route has been reconstructed mainly from crustal conditions far removed from the upper Grand. Read more about it in chapter 4.

- *The Third Age: Late Glacial, Interlobate conditions.* The Quaternary Ice Age framed the next set of critical changes involving the gorge, especially the final years of the last Laurentide Ice Sheet. Until the late glacial period, perhaps between 15,000 and 17,000 years ago, there probably were not even hints of the gorge. Its singular development involved glacial conditions, especially in the emergence of an ice-free area known as the Ontario Island. Influential factors were the work of, and interactions among, ice lobes, glacial lakes, and meltwater spillways. Read more in chapter 5.

- *The Fourth Age: Primordial Forest.* While recovering from the Ice Age for most of the last 10,000 years, the Grand River basin became part of an immense forest. Over many millennia, a person might have walked from here to Florida or northern British Columbia without leaving the woods. Under the forest canopy, the gorges continued to develop, but were regulated by the dense tree cover, fallen branches, and beaver dams. One suspects that erosion acted at a much slower pace than in the late glacial phase, or, indeed, than it does today. Read more in chapters 5 and 6.

- *A Fifth Age: in the Anthropocene.* See chapter 9, pages 177–82.

Elora Falls in 1899
(see fig. 9.2 for full image)

Chapter Nine

People in the Gorge

In country such as Southern Ontario now is, a new environment has been created in which the river must become the agent of man as well as of nature.
—L.J. Chapman and D.F. Putnam (1984)[1]

FIRST PEOPLES

The memories and moods of a thousand rivers are woven into Canada's history. For millennia, First Nations people have been travelling along these rivers and settling beside them. Theirs is by far the longest period of human interaction with the landscapes of the Grand. Yet, in the gorge, and indeed most of the upper Grand Basin, their footprint has been light, largely invisible.

From other sources we know they were regular visitors and had sacred places here. As the glaciers withdrew, some enduring effects may have come with the extinction of large mammals, where Paleolithic hunters exploited them. Much later, substantial settlements, hunting grounds, and trade routes occupied the surrounding lands. There was some clearing of the woodland, including uses of fire. Sometime after 500 CE, the beginnings of settled agriculture are evident.[2] Thus far, archaeological and ethnographic materials regarding Indigenous peoples come almost entirely from the Grand

Basin below and outside of the gorge. In the gorge, substantial reported remains of a pre-colonial, Indigenous presence were a few hundred traditional shell or *wampum* beads. These were found in the late nineteenth century in small caves and recesses around the "Hole in the Rock" (*Site 12* and fig. 2.16, page 47). Apparently, some of the beads were exhibited in the museum at the Elora School, and others sent to the Royal Ontario Museum, but none have been recovered since.

Too little is known in detail of what Indigenous people did in the gorge. However, modern debates increasingly suggest a special relationship between Indigenous peoples and Earth, the land, and all living things. Reports emphasize practices of care, respect for, and harmony with nature. Inspiration is sought in them for a modern conservation ethic, in contrast to dangers of modern extractive uses. Consider what may come from such perspectives as "Place-Thought," which describes an Indigenous understanding of the world "based upon the premise that land is alive and thinking and that humans and non-humans derive agency through the extensions of these thoughts."[3]

In recent years, efforts to do justice to First Nations' history have been growing. There is a need for Indigenous voices in landscape studies. Better awareness demands that their own languages, imagery, and stories be heard. I am not qualified to address such matters but urge others who are to do so.

Here, I mainly address avenues connecting gorge natural history to the settler-colonial story, notably in relation to Earth science. It is a small part of the pivotal role geology has played in modern Canadian political economy and identity, as well as the landscape.[4] From the 1840s, Elora Gorge and some local residents played parts in these innovations in Earth history.

EARTH SCIENCE AT ELORA

Modern professional studies in the gorge area can be traced back to 1843.[5] This was the first year of work by the Geological

Survey of Canada (GSC), and led to a report by Alexander Murray. At that time he did not identify the gorge rock type, nor its distinctive qualities. He did describe them in his 1848 GSC report. In that year too, a distinguished New York paleontologist, James Hall, visited the region and identified the Guelph fossil clam, *Megalomus canadensis*. It was singled out as a characteristic fossil, later renamed as *Megalomoidea canadensis*, discussed in chapter 3. Following his further explorations as the provincial geologist, Murray reported that "This shell… was seen in greatest abundance at Galt [today's Cambridge, Ontario] and at Elora, on the Grand River."[6]

In 1861, Robert Bell first named the Guelph Formation, from rocks found in the vicinity of Galt and Guelph. As noted in chapter 3, this label was confirmed and adopted by the first director of the GSC, William Logan (1789–1875), in his book *Geology of Canada* (1863). Elora was mentioned in GSC Reports for 1850 (published 1852), 1863, 1861–65, 1901 and 1941.

To get something of the flavour of the geological work, its limited resources and unique findings, one can refer to T.C. Weston's 1899 memoir, *Reminiscences among the Rocks*.[7] He described the summer of 1867 as "confined to collecting fossils from the Guelph formation at Guelph, Galt, Elora, Hespeler, and other localities."[8] He recalled how the "Guelph formation has yielded a large fauna of fossils…many of which were new to science—now in the cases of the Dominion Geological Museum."[9]

Others who found the gorge notable for its vertical limestone cliffs as well as fossils were not geologists, but helped them and championed their work.

From Blacksmith to Schoolteacher and Museum Curator

David Boyle (1842–1911) is one of the foremost public figures whose words and actions concerning the rocks and archaeology of the gorge have survived (fig. 9.1). He came to Canada from Scotland and lived in the Elora area through the 1860s and 1870s. Before this, he had trained as blacksmith. By the time he moved on, he had retired as Elora School's headmaster, thereafter gaining prominence as an archaeologist.

Through Boyle, geology played a special part in the education of village children. As a teacher he followed the theories of the Swiss educational reformer Johann Pestalozzi (1746–1827). He advocated for child-centred methods and learn-

Figure 9.1   **David Boyle, by John R. Connon. Courtesy of Archival & Special Collections, University of Guelph, Connon Collection, XR1 MS A114303.**

ing that develops the whole person, aided by observation of nature, cultural advancement, and acquisition of mental and practical skills. On the educational value of the gorge he said, "I can conceive of no more profitable and pleasing way for a boy to spend a few hours, than in such a place, where almost every few minutes the hammer exposes a gracefully formed shell."[10]

Boyle was also an early convert to Darwinism and brought it into his classroom. He shared his new fascination with rocks and evolutionary theory. He placed local findings into the context of international developments, and bridged specialized and popular interest—an effort the present book aspires to continue.

There is a very readable biography of Boyle by Gerald Killan, which also gives a picture of the society in which he moved and the people who shared his thoughts on geology and archaeological discoveries.[11] Boyle's own first publication was *On the Local Geology of Elora* (1875).[12] It began as a talk he delivered to the Elora Natural History Society in 1874, which he also helped found. His work conveyed a sense of the ancient, teeming, and exotic life in the Guelph Sea being rediscovered in the 1870s. Local amateur naturalists were delighted to discover that the rocks and fossils they found could be linked to intellectual developments in the big cities of Europe and North America. This had a direct bearing on the new theories of evolution and Earth science, helping, for instance, to establish evidence of the Silurian and its global significance.

In those early days, any careful or lucky search in the gorge was likely to unearth a previously unidentified fossil. A compelling illustration of this reality is the number and diversity of finds named after local personalities and sites—including the indefatigable Mr. Boyle, reflected in species names terminating in "*boylei*" see Selected Guelph Formation fossils, page 174.

Selected Guelph Formation fossils named for locations in Southern Ontario, or individuals (*) who worked in the Gorge area

Class Stromatoporoidea
Stromatopora galtensis
Stromatoporella elora
Stromatoporella elora minuta
Hermatostroma guelphica
Labechia durhamensis

Class Anthozoa
Favosites niagrensis
Fletcheria guelphensis

Phylum Brachiopoda
Monomorella durhamensis
Rhinobulus galtensis
Eospirifer niagrensis

Class Bivalvia
Megalomoides canadensis
Iliona canadensis
Prolucina galtensis

Class Gastropoda
Archinacella canadensis
Pleurotomaria townsendi*
Euomphalus galtensis

Euomphalopteris elora
Lophospira hespelerensis
Lophospira guelphica
Lophospira conradi*
Lophospira elora
Eotomaria durhamensis
Euotomaria galtensis

Hormotoma whiteavesi
Turritoma boylei*
Poleumita durhamensis
Holopea guelphensis
Diaphorostoma niagrense

Class Cephalopoda
Ascoceras townsendi*
Orthoceras brucensis

Class Ostracoda
Leperditia phaseolus guelphica
Leperditia balthica guelphica

Subclass Eurypterida
Eurypterus boylei*

Based on the Royal Ontario Museum's fossil collection. Courtesy of Janet Waddington, Paleontology Department, Royal Ontario Museum.

Boyle also established a library in Elora, part of the Mechanics Institute. It would be passed on to the village and became part of the local public library. When my family settled in Elora in the 1970s, I was amazed to find so many once Earth-shattering books in a small village library. They included Darwin's main works and many other classics of nineteenth-century scientific advancement. As late as the 1990s, when it became part of the Wellington County Library, the collection still contained some of Boyle's legacy. Of course, controversy swirled around the whole endeavour. The struggle between the Creation story in Genesis and Darwin's "heresy" was the subject of thundering sermons in local churches. Boyle had to tread carefully, but he became part of what Killan refers to as "Elora's Intellectual Awakening."

An Awakening

In 1874 the local *Lightning Express* newspaper reported:

> Elora sometime since started a public museum.... To give some idea of the curiosities of nature presented to Elorians, we have only to name *echinus*, a *Megalomus compressus,* specimens of *Heliophylum Halli, Zaphrentus prolifica, Orthis livia, Spirifer, Mediolopsis modiolaris, Ambatichia radiata, Orthoceras, Stenopora fibrosa* and *echinoid erecta*. Such trifles may do to begin with, but of course, something more rare must be procured to keep up any permanent interest in the "show."[13]

The reporter, tongue in cheek, used these words to introduce Boyle's "museum," where many of Elora's Guelph Formation fossils were first collected, classified, and exhibited. Boyle would have been among the first to enjoy the reporter's humour. Nevertheless, many of those jaw-cracking classical names refer to the best treasures in his collection; fossils that he or his pupils and friends had gathered.

Fossils exert a special fascination. As shown in chapter 3, they serve as keys to interpreting the bedrock of the gorge, and the erosional forms that are found there. Boyle encouraged his students at Elora School to look out for new fossils and archaeological finds. He founded his museum at the school in 1873. The students and various local enthusiasts added exhibits. Boyle passed on many of their finds to the GSC.

In 1875, the Dominion Geological Museum in Montreal sent Boyle a collection of rocks and fossils, carefully catalogued. He displayed them in the School Museum, which grew substantially over the decade of his work there. By the 1870s, according to schoolmaster Boyle, "Almost all children in the village upward of five or six years of age, can distinguish *Megalomus Canadensis,* quite as readily as they can an Early Rose potato or a Swedish turnip."[14] By the 1880s, local promotion, regular cleanups, and the arrival of a railway connection led the *Lightning Express* to announce that "excursions from neighbouring villages to the rocks and exquisite natural scenery at Elora are of weekly occurrence lately."[15]

The Guelph dolostone had some useful properties too. Early on, settlers used it as building material. There was a good deal of lime burning in local quarries to produce quicklime, used in the production cement, among other things. A few geologists and local enthusiasts contributed, unsuccessfully, to a search for oil. Southwest of Elora, petroleum was discovered at depth in Guelph Formation rocks. These were among the first deposits to be exploited in North America. Elora itself had a brief flurry of "oil fever." You can find some oily surfaces in the Guelph rock, and an oil-like odour when you break open some hand samples. However, oil was never found in commercial quantities in or near the gorge.

Professional and commercial interest turned, above all, to the large and lucrative aggregate resources beyond the gorge, mainly sand and gravel left by the Laurentide Ice Sheet. Interest in the gorge itself shifted to the picturesque scenery and its tourism potential. The story of the gorge was taken up and driven by journalists and local enthusiasts, self-trained in geology. They played a prominent part in championing the value of the gorge itself.

There were volunteer projects to clean up the gorge, whereas previously settlers were more likely to use it for activities like washing wool, operating lime kilns, and dumping garbage—not to mention clearing the trees. With greater interest and campaigns to attract visitors, increasing numbers arrived in summer. The arrival of the Credit Valley Railway in 1880 further encouraged tourism. As now, the lower Irvine became a place where families enjoyed picnics, explored, and waded in the stream. It used to be a favourite spot for wedding photographs. By 1878 the local newspaper carried promotional material and advertising to support positive views of the region's geology and landscapes, for instance, proclaiming the presence of:

Yawning chasms—Cliffs and Rocks
Rapids and Falls—Pools and Caves
Rare Archaeological and Botanical Specimens.[16]

INTO THE ANTHROPOCENE

Had there been no river, offering the promise of water-power for driving mills, and with its tributary streams watering the nearby valleys, there would have been no village and the founders would have looked elsewhere for a townsite.

—Hugh Templin (1933)[17]

The Falls at Elora were introduced at the beginning of this book as features separating two very different worlds: one a tribute to natural processes, the other to humans. This book has mainly considered the former; especially how river, slope, and chemical processes have fashioned a landscape in dolostone bedrock. Upstream of the Falls, however, and much more widely, human activity prevails (fig. 9.2). Where one might look for the river's activity between Elora and Belwood Lake, the scene has been shaped much more by industries, public works, housing, and the needs of shopping and agriculture.

Figure 9.2   **Elora Falls, Photographed in 1899, by John R. Connon. Courtesy of Archival & Special Collections, University of Guelph, Connon Collection, XR1 MS A114290.**

Overwhelmingly, this reflects some 180 years of European settlement and industry. And it is a context that brings special challenges for the study of landscapes. Again, the gorge emerges as an exception. It presents a series of landscapes mainly reflecting natural processes, but in the midst of a river basin and riverscapes almost completely dominated by human activity.

Meanwhile, notions of geological conditions have been changing quite dramatically, even since 1995 and the first edition of this book. Major challenges arise in relation to human societies as geological agents. Relevant proposals urge treating the latest geological time frame as the *Anthropocene* or "Human Epoch."[18] Of special concern is the extent to which the Anthropocene presents wider threats to the ecosphere, if not human survival.

Take the opportunity to scan the Grand River basin on a satellite image using a free online map platform; an appropriately Anthropocene technology! Scan the triangle bounded by Waterloo, Guelph, and Brantford. The marks of human occupancy and infrastructure carve up the space. In the various cities, the built environment is more extensive than the river channels and even the flood plains. That is true vertically as well as horizontally. The river here makes its way through artificial rectangular blocks and subdivisions: "cityscapes," some much higher than the gorge walls. It is continuously built up, fully urbanized.

Discussion of the Anthropocene tends to foreground concerns and changes that emerged after 1945—the time of the so-called Great Acceleration.[19] It has been marked by exponential growth of artificially built or excavated landforms, massive resource extraction, climate change, and the spread of toxic and synthetic materials. It includes the ultimate ecocidal threats from nuclear and biological weapons. Nowhere is the Great Acceleration more apparent and critical than in Southern Ontario, Canada's main engine of economic and environmental change.[20]

The Anthropocene comes freighted with fear for, dangers to, and the fragility of landscapes such as the gorge. Discussion of these issues must address serious challenges to the protection of living systems, including human heritage. In exploiting natural conditions, Europeans reshaped, removed, drowned, or damaged what had been here. Such is the evidence marshalled by those who argue that humans have come to exert a dominant influence on global geological change and ecosystems. But how can this be set within the portrait of the gorge developed so far?

Drainage Fragmentation

In the Anthropocene, what has happened to most rivers is termed **drainage fragmentation**.[21] Human impacts on rivers like the Grand involve, above all, disruptions of stream flows and sediment movement. The connectivity of streams tends to be disturbed, along with relationships of tributaries and main streams. The most fragile parts of riparian ecology are put at severe risk.[22]

Causes of fragmentation may involve dams, weirs, and other artificial river-training works. The latter include embankments, levees, locks, spurs, dykes, and concrete sills—as can be seen on the Grand through Elora and Fergus (fig. 9.3). Stream reaches are broken up and continuity interrupted by dumping and water extraction, by channelization, bank protections and revetments, channel lining or riprap armouring. Partial obstructions can impede, divert, or split river channels, upset the stability of the bed, or affect the types of sediment that accumulate.

In the world of rivers, drainage fragmentation is a basic aspect of the conversion of Canada to a late-modern, urban-industrial economy. The growth of Ontario towns has been pivotal to fragmentation, in particular at lakeshores and river crossings, and where waterways join and open out into one or more lakes.

Fragmentation has been the most drastic during the Great Acceleration. Global estimates suggest fragmented drainage affects 77 percent of the flow of rivers in the northern third of the Earth's land mass. No large rivers and few small ones

Figure 9.3　**The Grand River above the gorge and through Fergus fully illustrates drainage fragmentation and a riverscape in the Anthropocene.**

remain unaffected. They represent artificial disturbances throughout the river systems of Eurasia and North America.[23] Ontario is fully implicated, its rivers heavily administered, exploited, and disrupted. Adverse ecological impacts are typically contrasted with "natural rivers."[24]

Once more, Elora Gorge itself can seem out of place. Here, even in the Anthropocene, rivers in rock continue to respond mainly to natural processes, to drainage shaped by winter ice, freeze-thaw, flood-prone stream flows, and soluble stone. As I have already shown, natural erosion has dominated the record. In one respect the gorge survives by being carved in rugged, inhospitable rock that has tended to keep people at a distance.

Environmental damage has been reported—and also bemoaned—for most of Elora's modern history. As early as 1877 the editor of the *Lightning Express* complained: "if the work of spoliation and enclosure goes on a little longer unchecked, Elora's natural beauties will be worth less than nothing...the best view of the Falls is already cut off from public view."[25]

It would, however, be misleading to suggest humans are not integrally involved in the present state and likely fate of the gorge, thanks especially to a measure of neglect. This neglect was behind the return of tree cover early in the last century. An even narrower escape occurred in the 1970s. It involved plans to dam the Grand River downstream of Elora near West Montrose for flood control and to counter low flow conditions. Had the dam gone through, the gorge and much of the conservation area would now be a kind of artificially buried bedrock valley (BBV): a canyon filled with fetid water, municipal effluent, and mud.[26] Almost all the features introduced in this book would be drowned. In sum, no Elora Gorge!

Instead, thanks to protests and financial problems, the gorge remains a fragment of the "wild" in the midst of artifice. It is by no means entirely protected, but remains perennially at risk of being despoiled or converted to a "human-made" landscape.

These issues of the gorge also challenge common but misleading dichotomies and stereotypes like "nature versus humanity." Everywhere that humans go there is nature. Everything they do is constrained by natural laws and planetary systems; by gravity, solar energy, photosynthesis, wind, and wave. These are intertwined, inseparable features of human existence. Meanwhile, humans in the Anthropocene have developed the capacity to protect—and destroy—all parts of the ecosphere. Landscapes raise natural and cultural challenges together, not least questions of stewardship, heritage, and protection of life worlds. Even in a cynical age, people increasingly emphasize how survival demands more sensitive relations to Earth habitats—driven by or requiring values and ethics.

HERITAGE AND LANDSCAPES

Meanwhile, the human role in transforming the Earth's surface has been a source of great pride and wealth for some. The bridges and tunnels, mines and dams, grain fields and cities that surround the gorge are seen as signal achievements, symbols of Canada's prosperity. Yet, wherever these and other aspects of the Anthropocene are encountered, they involve adverse environmental impacts and ecological damage. There have been intermittent, sometimes vigorous, efforts at environmental protection. The Conservation Authority (GRCA) was established in 1948 to manage the watershed for human benefit. In 1994, the Grand River was designated as a Canadian Heritage River. Ongoing economic development and climate change are among major challenges for any individual or institution striving to protect this heritage.

For some years now, Earth science and landform studies have turned increasingly to identify and expound the merits of special sites, notably landscapes. This is an interest identified with *geosites*, *geoheritage*, and with related aspects of environmental conservation.[27]

In 1994 UNESCO (the United Nations Educational, Scientific and Cultural Organization) and the IUGS (International Union of Geological Sciences) launched a Global Geosites Project. A *geosite* may present unusual or definitive rock outcrops; striking landforms and their development. It may provide worthwhile teaching moments for features that are widespread. The Geological Society of America identifies three types of *geoheritage* sites:

Scientifically and educationally significant geoheritage sites include those with textbook geologic features and landscapes, distinctive rock or mineral types, unique or unusual fossils, or other geologic characteristics that are significant to education and research. *Culturally*

Figure 9.4   "Cave view" by David Wilcox (1945–2022). The photograph presents a striking view of the lower Irvine Creek and many of the key features of Elora Gorge. Dave is sadly missed as friend, photographer, musician, hands-on polymath, and resident of Elora. Reproduced with permission of the estate of David Wilcox.

significant geoheritage sites are places where geologic features or landscapes played a role in cultural or historical events. *Aesthetically significant geoheritage sites* include landscapes that are visually appealing because of their geologic features or processes.[28]

Elora Gorge presents something of all the above (fig. 9.4). This book has sought to explore it as a distinctive regional terrain, with special geological and landscape features. It has exemplary scientific, educational, cultural, and aesthetic appeal, especially inspiring themes from the presence of "rivers in rock." Elsewhere, bedrock rivers include popular, protected, and spectacular landforms. Some overwhelm with their sheer size, including the Grand Canyon, Kali Gandaki Gorge in Nepal, Niagara Falls and Gorge, and Blyde River Canyon (Mpumalanga, South Africa). However, even relatively small and local sites, as at Elora, can provide compelling insights.[29]

Also well represented here are "*viewpoint geosites*," notably the "classic viewpoints" visited in chapter 2 (*Sites 1–3*). These locations allow unobstructed views of the landscape, instructive bedrock outcrops and structures, characteristic landforms, and possibly fossil exposures.

Conversely, there are notable tourist destinations that warrant treatment as heritage geosites. As at Elora, care and restoration of such sites are important.[30] In all such places, a commitment to environmental protection and respect can be made; this is needed for what remains to survive and be enjoyed later. However, this also favours a different sort of language than is common in Earth science. The *geosite* concept offers the benefits of, and a way to recognize and reconcile, the multiplicity of First Nations', settler community, artistic, and scientific perspectives on landscape.

Notes

PREFACE

1   Kenneth Hewitt, *Elora Gorge: A Visitor's Guide* (Toronto: Stoddart, 1995).

2   David Boyle, *On the Local Geology of Elora,* Selected Papers from Proceedings of the Elora Natural History Society, 1874–75 (Elora, ON: J. Townsend, 1875), https://archive.org/details/selectedpapers00enhsuoft/mode/2up. Boyle's special role in the human story of the gorge is summarized in chapter 9.

3   Gordon Nelson, ed., *Towards a Grand Sense of Place: Writings on Changing Environments, Land-Uses, Landscapes, Lifestyles and Planning of a Canadian Heritage River*, Waterloo Department of Geography Publications Series (Waterloo, ON: Heritage Resources Centre, University of Waterloo, 2004); Stephen Thorning, "Fish, clean rivers a concern in Wellington 165 years ago," *Wellington Advertiser*, April 19, 1999; repr. May 20, 2021, https://www.wellingtonadvertiser.com/fish-clean-rivers-a-concern-in-wellington-165-years-ago/; Thorning, "Conservation initiatives of 1913 failed to get action," *Wellington Advertiser*, September 22, 2006; repr. July 15, 2020, https://www.wellingtonadvertiser.com/conservation-initiatives-of-1913-failed-to-get-action/.

4   "Elora Gorge," Grand River Conservation Authority, accessed May 15, 2023, https://www.grandriver.ca/en/outdoor-recreation/Elora-Gorge.aspx.

5   For information on seasonal closures and wheelchair accessibility in the Elora Gorge Conservation Area, see Grand River Conservation Authority, "Elora Gorge."

1. INTRODUCTION: RIVERSCAPES IN ROCK

1   The origins of the bedrock and its qualities are addressed in chapter 3. How it responds to rivers in rock is addressed in chapter 6, the rock wall features in chapter 7, and chemical erosion in chapter 8.

2   For an overview of the basin and landforms of the Grand, a good place to begin is L.J. Chapman and D.F. Putnam, *The Physiography of Southern Ontario*, Ontario Geological Survey Special Volume 2, 3rd ed. (Toronto, ON: Ontario Ministry of Natural Resources, 1984). A broader, more detailed overview of the regional setting is provided in Nelson, *Towards a Grand Sense of Place.*

3   See Dwight Boyd et al., "The Water Regime or Hydrology of the Grand River Watershed," in Nelson, *Towards a Grand Sense of Place*, 47–58; Marie Sanderson and Brian Mills, "Climate Variability in the Grand River Watershed," in Nelson, *Towards a Grand Sense of Place*, 35–46.

4   Contemporary landscape-shaping forces are discussed in chapter 6.

5   John Robert Connon, *The Early History of Elora, Ontario and Vicinity*, reprinted with introduction by Gerald Noonan (1930; repr., Waterloo, ON: Wilfrid Laurier University Press, 1975).

6   Michael Henry and Peter Quinby, *Ontario's Old-Growth Forests*, 2nd ed. (Markham, ON: Fitzhenry & Whiteside, 2021).

7   The history of the gorge bedrock is discussed in chapter 3.

2. FIRST-HAND EXPERIENCE

1   Locally called "The Falls," but more strictly a cascade rather than a true waterfall in that it lacks vertical or overhung descent.

2   This is Elora's Victoria Crescent neighbourhood, which some advocates have campaigned to be recognized as a heritage district. Elora Victoria Crescent Neighbourhood Heritage Conservation District Study Subcommittee, *Elora's Victoria Crescent Neighbourhood Heritage Conservation District Study* (Elora, ON: Elora Victoria Crescent Neighbourhood Heritage Conservation District Study Subcommittee, 2010).

3   Connon, *Early History of Elora*, 171.

4   The discussion returns briefly to the human story and landforms indicative of the Anthropocene in chapter 9.

5   More features around the Falls are introduced in Excursion 2.

6   Stream bed erosion and other fluvial processes are discussed in chapter 6.

7   Different types of stream junctions are discussed in chapter 6, pp. 126–29.

8   Photographs and a discussion of the various bridges can be found in Roberta Allan, compiler, *History of Elora* (Elora, ON: Elora Women's Institute, 1982), 140–44.

9   Slope processes in the gorge are discussed in chapter 7.

10 Stress release fractures are discussed in chapter 7, pp. 141–46.

11 This rockfall episode is discussed in chapter 7, pp. 148–49.

12 Ongoing demolition and new construction may change details of access after this goes to press. Please check at the Elora and District Community Centre, 29 David Street, Elora.

13 Connon, *Early History of Elora*, 45.

14 The higher ones may even have started as glacial spillways, when the Ice Sheet covered much of the Upper Grand basin. See chapter 5.

3. BEDROCK

1   Connon, *Early History of Elora*, 44.

2   William E. Logan, *Geology of Canada: Geological Survey of Canada—Report of Progress from its Commencement to 1863* (Montreal: Dawson Bros., 1863).

3   As is evident from fig. 3.2, Eras are subdivided into Periods
which may, in turn, be divided into upper, middle, and lower.
Those like the Guelph may subdivided into Formations, and
further into Members. These may have different names in
different places, usually based on a type-location. Recently,
within the Guelph Formation near Elora, a lower *Wellington Member* and an upper *Hanlon Member* have been dif-
ferentiated. F.R. Brunton, "Project Unit 08-004. Update of
Revisions to the Early Silurian Stratigraphy of the Niagara
Escarpment: Integration of Sequence Stratigraphy, Sedimentology and Hydrogeology to Delineate Hydrogeologic
Units," in *Summary of Field Work and Other Activities 2009*,
Ontario Geological Survey, Open File Report 6240 (Sudbury,
ON: Ministry of Northern Development and Mines, 2009),
25-1–25-20.

4   Nick Eyles, *Ontario Rocks: Three Billion Years of Environmental Change* (Markham, ON: Fitzhenry & Whiteside,
2002).

5   Boyle, *On the Local Geology of Elora*, 5.

6   Candace Brintnell, "Architecture and Stratigraphy of the
Lower Silurian Guelph Formation, Lockport Group, Southern Ontario and Michigan" (MSc thesis, University of Western Ontario, 2012), https://ir.lib.uwo.ca/etd/632.

7   B.V. Sanford, "Silurian of Southwestern Ontario," in Ontario
Petroleum Institute, Proceedings of the 8th Annual Conference, Technical Paper No. 5, 1–44 (London, ON: Ontario
Petroleum Institute, 1969); Brintnell, "Architecture and Stratigraphy," 74.

8   F.R. Brunton et al., "Silurian Reef Episodes, Changing Sea-
scapes and Paleobiogeography," in *Silurian Cycles: Linkages
of Dynamic Stratigraphy with Atmospheric, Oceanic, and
Tectonic Changes*, Proceedings of the Second International
Symposium on the Silurian System, New York State Museum
Bulletin 491 (Albany, NY: State University of New York, State
Education Department, 1998), 259–76.

9   Brintnell, "Architecture and Stratigraphy," 74.

10   Work on them has advanced considerably in the present century, but most of it remains in technical and government documents. A good sense of Guelph Formation discoveries can be found in Maurice E. Tucker and V. Paul Wright, *Carbonate Sedimentology* (Malden, MA: Blackwell Scientific, 1990); Mario Coniglio, Qing Zheng, and Terry R. Carter, "Dolomitization and Recrystallization of Middle Silurian Reefs and Platformal Carbonates of the Guelph Formation, Michigan Basin, Southwestern Ontario," *Bulletin of Canadian Petroleum Geology* 51, no. 2 (June 2003): 177–99, https://doi.org/10.2113/51.2.177; Brintnell, "Architecture and Stratigraphy"; and Bob Janzen, *Geology of the Grand River Watershed: An Overview of Bedrock and Quaternary Geological Interpretations in the Grand River Watershed* (Cambridge, ON: Grand River Conservation Authority, 2018). Some disagreement persists. A few scientists would still like to see the Stroms placed in or closer to the corals.

11   Some scientists claim recently discovered sponges in the deep ocean are surviving Stroms.

12   W.A. Parks, *The Stromatoporoids of the Guelph Formation in Ontario*, University of Toronto Studies, Geological series, No. 4 (Toronto, ON: University of Toronto, 1907), https://archive.org/details/cihm_80863.

13   A simple way to determine whether a sample of limestone is dolomite or not requires a few drops of dilute (10 percent) hydrochloric acid. If the material is dolomite, the liquid will spread over the rock without discernable reaction. If it is calcium carbonate or aragonite, the acid will "fizz" profusely. This means that, if you have found an uncontaminated surface or intact sample of dolomite bedrock, there will be no reaction. If there is, instead, a strong reaction, you have found calcium carbonate. Expect to find both at many exposures in the gorges. The main body of the rock is dolomitic, interspersed by later deposits or films of calcium carbonate

precipitated over the surface or at springs. Some prefer to reserve the term dolomite for the mineral and call the rock dolostone.

14   Coniglio, Zheng, and Carter, "Dolomitization and Recrystallization."

4. TECTONICS AND ANCIENT GEOGRAPHIES

1   The first, the Cambrian Period, is missing at the surface in Southern Ontario but found at depth.

2   Other relevant geological structures are introduced below.

3   Carl K. Seyfert, ed., *Encyclopedia of Structural Geology and Plate Tectonics* (New York: Van Nostrand Reinhold, 1987).

4   Discussion of plate tectonics is widely available in print and online. Comprehensive coverage of matters raised here is available in Seyfert, *Encyclopedia of Structural Geology.*

5   Seyfert, 83 and 313. The original Gondwana is in what is now central India, by coincidence about 300 kilometres east-southeast of the Ellura caves, a world heritage site from which Elora's name is derived.

6   Seyfert, 89.

7   Seyfert, 67.

8   Seyfert, 217.

9   Seyfert, 145.

10   Seyfert, 95.

11   Seyfert, 96.

5. ICE AGE ORIGINS OF THE GORGE

1   *Lightning Express* (Elora, ON), April 22, 1875, quoted in Hewitt, *Elora Gorge*, 8.

2   Boyle, *Local Geology of Elora*, 4.

3   Getting its name from the same source as the North American Craton, or *Laurentia,* met in chapter 4.

4   P.F. Karrow, "Quaternary Geology of the Great Lakes Sub-region," in *Quaternary Geology of Canada and Greenland*, Geology of Canada No. 1, ed. R.J. Fulton (Ottawa, ON: Geological Survey of Canada, 1989), 326–50; Alan V. Morgan and Paul F. Karrow, "Bedrock Geology of the Grand River Watershed," in Nelson, *Towards a Grand Sense of Place*, 11–20.

5   Chapman and Putnam, *Physiography of Southern Ontario*.

6   Chapman and Putnam established the layout, growth, and disappearance of the various ice lobes, interlobate environments, and their relations to moraines and former glacial lakes. Above all, for the region around the gorge, there is their painstaking mapping of the Ontario Island: Chapman and Putnam, 26–33. Arguably, reconstruction of these details is amongst their most amazing and significant work.

7   Alan E. Kehew et al., "Proglacial Megaflooding along the Margins of the Laurentide Ice Sheet," in *Megaflooding on Earth and Mars*, ed. Devon M. Burr, Paul A. Carling, and Victor R. Baker (Cambridge: Cambridge University Press, 2009), 104–27; Mandy J. Munro-Stasiuk et al., "The Morphology and Sedimentology of Landforms Created by Subglacial Megafloods," in Burr et al., *Megaflooding on Earth and Mars*, 78–103.

8   See Virgil Martin, "Grand River Vegetation and Wildlife," in Nelson, *Towards a Grand Sense of Place*, 59–83.

9   Woody debris is discussed further in chapter 6, pp. 120–21.

6. RIVERS IN ROCK

1   See Ellen E. Wohl, "Bedrock Channel Morphology in Relation to Erosional Processes," in *Rivers over Rock: Fluvial Processes in Bedrock Channels*, Geophysical Monograph 107, ed. Keith J. Tinkler and Ellen E. Wohl (Washington, DC: American Geophysical Union, 1998), 133–51.

2   Chapman and Putnam, *Physiography of Southern Ontario*, 144.

3   Phillips and Desloges provide impressive coverage of alluvial streams and related postglacial Ontario landforms, but mention bedrock streams only in passing without elaboration. R.T.J. Phillips and J.R. Desloges, "Glacial Legacy Effects on River Landforms of the Southern Laurentian Great Lakes," *Journal of Great Lakes Research* 41, no. 4 (December 2015): 951–64.

4   Ellen Wohl, *Mountain Rivers Revisited*, Water Resources Monograph 19 (Washington, DC: American Geophysical Union, 2010), 81.

5   Paul A. Carling et al., "The Bubble Bursts for Cavitation in Natural Rivers: Laboratory Experiments Reveal Minor Role in Bedrock Erosion," *Earth Surface Processes and Landforms* 42, no. 9 (July 2017): 1308–16.

6   Nurit Shtober-Zisu, Hani Amasha, and Amos Frumkin, "Inland Notches: Implications for Subaerial Formation of Karstic Landforms—An Example from the Carbonate Slopes of Mt. Carmel, Israel," *Geomorphology* 229 (January 2015): 85–99.

7   I.P. Martini, "Gravelly Flood Deposits of Irvine Creek, Ontario, Canada," *Sedimentology* 24, no. 5 (October 1977), 603–22.

8   Michael Church, "Bed Material Transport and the Morphology of Alluvial River Channels," *Annual Review of Earth and Planetary Sciences* 34 (May 2006): 325–54.

9   Janet H. Curran and Ellen E. Wohl, "Large Woody Debris and Flow Resistance in Step-Pool Channels, Cascade Range, Washington," *Geomorphology* 51, nos. 1–3 (March 2003): 141–57.

10   The term "discordant drainage" is also used to describe stream patterns that are at odds with geological structure.

11   Wohl, *Mountain Rivers Revisited*, 2–3.

12   The period of interlobate environments in the gorge area is discussed in chapter 5, pp. 94–101.

7. ROCK WALLS: THE STRONG AND THE WEAK

1   M.J. Selby, *Hillslope Materials and Processes*, 2nd ed. (Oxford: Oxford University Press, 1993), 333–38.

2   Edwin L. Harp and Randall W. Jibson, "Anomalous Concentrations of Seismically Triggered Rock Falls in Pacoima Canyon: Are They Caused by Highly Susceptible Slopes or Local Amplification of Seismic Shaking?" *Bulletin of the Seismological Society of America* 92, no. 8 (December 2002): 3180–89.

8. KARST

1   Derek C. Ford and Paul Williams, *Karst Geomorphology and Hydrology*, rev. ed. (Hoboken, NJ: Wiley, 2007).

2   Brunton and Dodge map the Guelph Formation dolostone in the gorge area as "potential karst," rock considered susceptible to karstification but as yet lacking notable developments. F.R. Brunton and J.E.P. Dodge, *Karst of Southern Ontario, Including Manitoulin Island*, Ontario Geological Survey, Groundwater Resources Study 5 (Sudbury, ON: Ontario Ministry of Northern Development and Mines, 2008).

3   J.N. Jennings, *Karst Geomorphology*, 2nd ed. (New York: Blackwell, 1985), 73.

4   A spelunker is someone who explores caves. Speleology is the science of caves.

5   The Guelph Formation karst at nearby Rockwood has received more attention. There are major contributions to global karst science by Canadians, notably Derek C. Ford and his team at McMaster University (for example, Ford and Williams, *Karst Geomorphology and Hydrology*).

6   Marie Sanderson, *The Grand Climate: Weather and Water in the Grand River Basin* (Cambridge, ON: Grand River Foundation, 1998); Sanderson and Mills, "Climate Variability."

7   Human activity in the Grand River watershed is discussed in chapter 9.

8   Keith J. Tinkler, "Fluvially Sculpted Rock Bedforms in Twenty Mile Creek, Niagara Peninsula, Ontario," *Canadian Journal of Earth Sciences* 30, no. 5 (May 1993): 945–53; Tinkler and Wohl, *Rivers over Rock.*

9   J.P. Greenhouse and P.F. Karrow, "Geological and Geophysical Studies of Buried Valleys and their Fills near Elora and Rockwood, Ontario," *Canadian Journal of Earth Sciences* 31, no. 12 (December 1994): 1838–48; Janzen, *Geology of the Grand River Watershed*, 4.

10 Jason Cole, Mario Coniglio, and Simon Gautrey, "The Role of Buried Bedrock Valleys on the Development of Karstic Aquifers in Flat-Lying Carbonate Bedrock: Insights from Guelph, Ontario, Canada," *Hydrogeology Journal* 17, no. 6 (September 2009): 1411–25; Colby M. Steelman et al., "Geophysical, Geological, and Hydrogeological Characterization of a Tributary Buried Bedrock Valley in Southern Ontario," *Canadian Journal of Earth Sciences* 55, no. 7 (July 2018): 641–58.

11 Cunhai Gao, "Buried Bedrock Valleys and Glacial and Subglacial Meltwater Erosion in Southern Ontario, Canada," *Canadian Journal of Earth Sciences* 48, no. 5 (May 2011): 801–18.

12 Cole, Coniglio, and Gautrey, "The Role of Buried Bedrock Valleys."

13 Nicholas Eyles et al., "Bedrock Jointing and Geomorphology in Southwestern Ontario, Canada: An Example of Tectonic Predesign," *Geomorphology* 19, nos. 1–2 (May 1997): 17–34.

14 Gao, "Buried Bedrock Valleys."

15 A compelling sense of these differences can be seen in the mapping of Pleistocene geology: P.F. Karrow, *Pleistocene Geology of the Guelph Area, Southern Ontario*, Geological Report 61 (Toronto, ON: Ontario Department of Mines, 1968).

9. PEOPLE IN THE GORGE

1   Chapman and Putnam, *Physiography of Southern Ontario*, 111.

2   Gary Warrick, "Early Aboriginal Occupation of the Grand River Watershed." In Nelson, *Towards a Grand Sense of Place*, 83–95.

3   Vanessa Watts, "Indigenous Place-Thought and Agency amongst Humans and Non-Humans (First Woman and Sky Woman Go on a European World Tour!)," *Decolonization: Indigeneity, Education and Society* 2, no. 1 (May 2013): 20–34, 21.

4   As convincingly shown by Suzanne Zeller in *Inventing Canada: Early Victorian Science and the Idea of a Transcontinental Nation* (Toronto: University of Toronto Press, 1987).

5   I am indebted, in particular, to archival research by Julia Roberts, under the supervision of Wilfrid Laurier University history professor Dr. Suzanne Zeller.

6   Alexander Murray, "Report of Alex. Murray, Esq., Assistant Provincial Geologist, addressed to W. E. Logan, Esq., Provincial Geologist," in *Geological Survey of Canada: Report of Progress for the Year 1850–51*, ed. William E. Logan (Quebec: John Lovell, 1852), 13–33, 25.

7   Thomas Chesmer Weston, *Reminiscences Among the Rocks in Connection with the Geological Survey of Canada* (Toronto, ON: Warwick Bros & Rutter, 1899).

8   Weston, 49–50.

9   Weston, 50.

10 Boyle, *Local Geology of Elora*, 5. Or girl, I can say, after the many hours my daughters spent here.

11 Gerald Killan, *David Boyle: From Artisan to Archaeologist* (Toronto: University of Toronto Press, 1983).

12 Boyle, *Local Geology of Elora*.

13 *Lightning Express* (Elora, ON), June 12, 1874, quoted in Hewitt, *Elora Gorge*, 58.

14 Boyle, 6.

15 "Local News," *Lightning Express* (Elora, ON), September 9, 1880.

16 "Do Something!" *Lightning Express* (Elora, ON), May 17, 1877.

17 Hugh Templin, *Fergus: The Story of a Little Town* (Fergus, ON: Fergus News-Record, 1933), 15.

18 Paul J. Crutzen, "The 'Anthropocene,'" in *Earth System Science in the Anthropocene: Emerging Issues and Problems*, ed. Eckart Ehlers and Thomas Krafft (Berlin: Springer, 2006), 13–18; Sophie Hackett, Andrea Kunard, and Urs Stahel, eds., *Anthropocene: Burtynsky, Baichwal, de Pencier* (Toronto: Art Gallery of Ontario, 2018) exhibition catalogue.

19 J.R. McNeill and Peter Engelke, *The Great Acceleration: An Environmental History of the Anthropocene since 1945* (Cambridge, MA: Harvard University Press, 2014).

20 Kenneth Hewitt, "Human Society as a Geological Agent," in *Quaternary Geology of Canada and Greenland*, Geology of Canada No.1, ed. R.J. Fulton (Ottawa, ON: Geological Survey of Canada, 1989), 624–34.

21 Will Steffen, Paul J. Crutzen, and John R. McNeill, "The Anthropocene: Are Humans Now Overwhelming the Great Forces of Nature?" *Ambio* 36, no. 8 (December 2007): 614–21.

22 Ellen E. Wohl et al., "The Natural Sediment Regime in Rivers: Broadening the Foundation for Ecosystem Management," *BioScience* 65, no. 4 (April 2015): 358–71.

23 Mats Dynesius and Christer Nilsson, "Fragmentation and Flow Regulation of River Systems in the Northern Third of the World," *Science* 266, no. 5186 (November 1994): 753–62.

24 Christer Nilsson, Catherine A. Reidy, Mats Dynesius, and Carmen Revenga, "Fragmentation and Flow Regulation of the World's Large River Systems," *Science* 308, no. 5720 (April 2005): 405–8; Elizabeth P. Anderson, Catherine M.

Pringle, and Mary C. Freeman, "Quantifying the Extent of River Fragmentation by Hydropower Dams in the Sarapiquí River Basin, Costa Rica," *Aquatic Conservation: Marine and Freshwater Ecosystems* 18, no. 4 (June 2008): 408–17.

25   "Do Something!" *Lightning Express* (Elora, ON), May 17, 1877.

26   Buried bedrock valleys are discussed in Chapter 8, pp. 163–65.

27   Luis Carcavilla, Juan José Durán, Ángel García-Cortés, and Jerónimo López-Martínez, "Geological Heritage and Geo-conservation in Spain: Past, Present, and Future," *Geoheritage* 1 (October 2009): 75–91.

28   Geological Society of America, *GSA Position Statement: Geoheritage* (Boulder, CO: Geological Society of America, 2017), https://www.geosociety.org/gsa/positions/position20.aspx.

29   Ellen E. Wohl, "Spatial Heterogeneity as a Component of River Geomorphic Complexity," *Progress in Physical Geography* 40, no. 4 (August 2016): 598–615.

30   José Brilha and Emmanuel Reynard, "Geoheritage and Geo-conservation: The Challenges," in *Geoheritage: Assessment, Protection, and Management*, ed. Emmanuel Reynard and José Brilha, 433–38 (Amsterdam: Elsevier, 2018).

Glossary of Landform-Related Terms Applied to the Gorge

This Glossary defines terms employed in the text that relate to landforms and earth surface processes, and where relevant describes their significance at Elora Gorge. The terms are in **blue bold** where they first appear in the book.

Accordant junction: a frequently observed tendency of tributary streams to join the main stream at the same elevation. This applies to the confluence of Irvine Creek and the Grand River, but all others in the gorge are, by contrast, **discordant junctions**.

Alluvium: unconsolidated deposits and soil, usually composed of sand, silt and clay, carried by streams and deposited along their channels, in ponds and flood plains. Typically these are absent from Elora Gorge in contrast to surrounding valleys.

Anchor Ice: cold season ice frozen to the stream bed. When it is released, rock fragments are likely to be pulled away, playing a significant part in gorge erosion in winter and spring.

Aragonite: a less stable form of *calcium carbonate* ($CaCO_3$), mainly in shells and other organic remains. Tends to transform into **calcite**.

Bare karst: a solution feature exposed at the surface in outcrops of limestone bedrock. Most **karst** observed in the Elora Gorge is bare karst, as opposed to **mantled karst,** which applies to much of the surrounding limestone.

Barrier reef: an extensive offshore system of organic, tropical reefs in shallow water. In modern times, typically coral reef. In the Silurian rocks of Elora Gorge, the reef system was built predominantly by *stromatoporoids,* long-extinct sponges of the *phylum Porifera.*

Base level: the level to which a stream tends to flow and usually cannot erode lower—obviously, sea or lake level, or the height of the main stream—but also subject to modification by resistant material in the channel, mainly bedrock in the gorge.

Border ice: frozen margins of streams, usually the first to freeze in winter and likely to cause erosion along the banks of the gorge.

Buried bedrock valley (BBV): a gorge, typically discontinuous, cut during late- or postglacial conditions. Most seem to have been meltwater spillways, subsequently buried by glacial, *periglacial* and/or *fluvial* processes, for example the Elora Buried Valley. More than a dozen BBV sites have been identified in the Grand River basin. Some are major aquifers.

Calcareous: partly or mainly containing *calcium carbonate.* The rock of Elora Gorge is an example.

Calcite: a mineral composed of *calcium carbonate* ($CaCO_3$), the more stable and main component of limestone, including the gorge rocks (compare **aragonite** and **dolomite**).

Carbonate: any of the members of a group of minerals found mostly in limestone and **dolomite,** including **aragonite** and **calcite.** The term is also used to refer to rock composed predominantly of these minerals and is the preferred scientific term for limestone.

Cavitation: a mechanism by which vapour bubbles grow and collapse suddenly in low pressure regions of fast-flowing water. Sudden implosion pressures may be sufficient to pit and disintegrate rock.

Collision zone: in *plate tectonics*, a zone where boundaries of lithospheric crustal plates— oceanic, continental or both—converge, typically creating mountain belts, deep ocean trenches, or volcanic arcs.

"Cornice rock": term coined by the author to describe a cliff capped by slabs or other rectilinear forms overhanging the gorge (compare **potter's rock**).

Craton: a massive and resistant continental crust, e.g., the *North American Craton*.

Crevice cave: a tunnel, gap, or chasm associated with deep tensile fractures opening out behind cliff faces. Widely present on the Niagara Escarpment, they play an important role along the walls of Elora Gorge and are a major factor in its appearance and evolution.

Discordant junction: a confluence where a tributary joins the main stream in a steep fall, erosion having failed to keep pace with the main stream; for example, in the Grand Gorge, "The Cascade" (*Site 28*); see also **hanging tributary**.

Doline: see **sinkhole**.

Dolomite: a mineral composed mainly of *calcium magnesium carbonate* ($CaMg[CO_3]_2$), or sedimentary rock composed mainly of the same (the latter also called **dolostone**). Generally, chemically altered limestone deposits, as in Elora Gorge, formed by a process of *dolomitization* that occurs with the passage of fresh water and brines, leading to replacement of a large fraction of calcium ions in *calcium carbonate* (limestone) with magnesium ions.

Dolostone: relatively hard limestone with high proportion of **dolomite**, as in gorge rock.

Drainage fragmentation: disruption and damming of streams, including water extraction and diversion, toxic effluent and dumping, interference with riparian ecology and life forms. It is widely observed along the Grand River above and below the gorge.

Drift: deposits, mainly Quaternary, formerly attributed to marine inundations on land, now to sediments of glacial origin, widespread in regions surrounding the gorge (see also **erratic**).

Dripstone: calcite or similar mineral deposit formed by precipitation of *calcium carbonate* from dripping water, usually in a cave, that also can form **stalactites** and **stalagmites**.

Erratic: a far-travelled stone, usually of different composition from surrounding ground or bedrock, largely used to refer to boulders carried and deposited by former glaciers. Scattered examples are widely observed in the gorge and throughout surrounding terrain.

Flowstone: a sheet-like deposit of *calcium carbonate*, typically on walls and floors of limestone caverns.

Fragmentation: see **drainage fragmentation**.

Frazil ice: a slushy mass of ice particles in a stream or pond. In the gorge it tends to form on cold, clear nights and may eventually sink to the bottom and freeze solid.

Geomorphology: the study of landforms and Earth surface conditions. Used interchangeably with *physiography*.

Glacial spillway: a stream channel, usually for meltwater, fed by ablation of glacier ice. The gorge probably originated as, or in, systems of spillways that opened up with the retreat of the *Laurentide Ice Sheet.*

Hanging cave: a cavern opening out in rock walls above present river level. There are many examples in Elora Gorge.

Hanging rock: a cliff projecting beyond the vertical and/or a lower cliff face, a prevailing feature of the gorge walls.

Hanging tributary (or discordant tributary): a tributary stream that reaches the main stream at a higher level and descends steeply to form a **discordant junction** (compare **hanging valley**). Examples are found throughout the gorge, unlike the **accordant junction** of the Irvine with the Grand, suggesting complex drainage evolution.

Hanging valley: a valley formed by a **hanging tributary**, where erosion has not kept pace, or caught up with the main stream. See also **discordant junction**.

Ice jam: a broken mass of river ice, usually blocked at constrictions or by obstructions along the river. Can create ice dams and break suddenly to cause an ice jam flood. Typically seen in Elora Gorge in spring at sites like David Street Bridge, and at the flats along the Irvine above Salem.

Interlobate environment: the outer region of an ice sheet where it splits into two or more lobes. It tends to create a complex terrain of **moraines**, glacial lakes, **glacial spillways,** and ice lobes which may interact. It is hypothesized that Elora Gorge originated in such terrain.

Karst: a landscape produced by dissolution and precipitation of *calcium carbonate* mainly in limestones and in the **dolostone** of the gorge.

Mantled karst: solution forms in limestone rock that are hidden under soil or superficial deposits; rare in Elora Gorge, prevalent in surrounding terrain.

Mass wasting: slope processes where erosional debris moves under the influence of gravity, and according to local moisture supply, freeze-thaw, chemical and physical weathering, or human disturbance. In the gorge it involves, especially, **rockslides** and **rockfalls**.

Megaflood: an exceptional deluge or outburst flood, notably involving a sudden release of meltwater impounded by ice or **moraine** around Quaternary ice sheets. Thought to have been critical in late glacial conditions around the Laurentide Ice Sheet when the gorge originated.

Moraine: a deposit of glacially transported debris. Limited amounts remain in the gorge, but are widely present in surrounding areas.

Natural stacking: an organized arrangement of stones or coarse *clasts* (pieces of rock), preferentially emplaced by flowing water or mass movement processes.

"Notch and visor": a cliff face landform with overhanging roof and deep recess or rock shelter, found repeatedly at the base of gorge cliffs.

Parting: joint, crack, or *bedding plane* that can constitute the main line of weakness in bedrock. There is a great diversity of partings in the gorge rocks, reflecting their original marine reef environments and subsequent history.

Periglacial regime: cold region environments and landforms dominated by ground ice, frost, and thaw processes, by snow and possibly **permafrost**. Found beyond the margins of glaciated areas and noted for "patterned ground" features generated by freeze-thaw processes. This affected the gorge early in the Holocene (the geological epoch that began after the last Ice Age).

Permafrost: ground that has been below 0° C for at least two consecutive years. Can become ice-free, but a variety of ice buildups and ice-cored features are important in landform features or due to intermittent melting and degradation. Likely present in the gorge terrain of the early postglacial centuries.

Pothole: basin-like, circular depression, ranging from fist size to metres in diameter, scoured by rotational eddies in streams. Many and varied examples can be found in the gorge.

"Potter's rock": term coined by the author to describe cliffs in **dolostone** salient along the gorge, capped by rounded forms resembling ceramic pots.

Primary erosion: Weathering and removal of bedrock.

Rockfall and **rockslide:** a sudden mass movement, usually from failures in bedrock. **Rockfalls** descend partly or wholly by freefall and a tumbling motion. **Rockslides** remain in continuous contact with the slope as they descend.

Rock shelter: a recession in a cliff face or at its base, with overhanging roof, possibly due to solution or frost weathering, or both. In the gorge there are hundreds of both ancient and recent origin.

Sessile organisms: creatures that inhabit the sea floor and may be attached to it or burrow into it. In the gorge rocks, they included reef-building corals and sponges.

Sharpstone: an angular rock fragment or broken *clast* with sharp edges and points, typical of **rockfall** fragments in the gorge.

Sheet ice: river ice that forms across the whole surface of the stream.

Sinkhole: a depression in limestone, also called a **doline**, due to solution or roof collapse, or both. Surface water drains underground, possibly into large cavern systems. There is a wide variety of forms in Elora Gorge, mostly small.

Speleothem: a general term for redeposited *calcium carbonate* in **karst** terrain and a great variety of forms including **stalactites** and **stalagmites**. There are countless minor examples in the gorge, mostly small.

Stalactite: a type of **speleothem** deposited from water saturated with *calcium carbonate*, common in limestone caverns. Minor examples are found in caves in the gorge.

Stalagmite: a type of **speleothem** formed of **dripstone** built by *calcium carbonate* precipitated from falling droplets, often below or paired with **stalactites**.

"Stone Sidewalk": a distinctive natural cliff feature with **hanging rock** in the gorge, aligned along old, abandoned gorge channels and resembling human-made footpaths or sidewalks.

Strath terrace: a riverbank carved in bedrock. Terrace cut by combined lateral and vertical stream incision. These are common along gorge flanks.

Superimposed drainage: a stream cutting down into substrate from differently configured valley form, rock structure, or substrate material. Elora Gorge reflects substantial superimposition of **glacial spillways** and streams into underlying **dolostone**.

Travertine: see **tufa**.

Tufa: a fresh water **carbonate** precipitated from *calcium carbonate*-saturated springs. An important source of **karst**

features in limestone terrain, usually precipitated in or over bacteria, algae, and mosses. It may be called "ambient temperature" or "cold" **travertine**. Travertine is more commonly associated with precipitated carbonate at hot springs. Active and dried up tufa deposits are common in the gorge.

Tunnel valley: a subglacial stream or spillway, usually at or near glacier margins, possibly flowing under pressure from surrounding ice and subject to outburst floods. Thought by some to have been a typical development of late glacial conditions and **glacial spillways** in the gorge region, perhaps associated with gorge initiation.

Unconformity: a surface generated by erosion marking and tracing the separation of older from younger rock.

Vug: a void or cavity in bedrock, characteristic of exposures of the **dolostone** bedrock in the gorge. The term originated among miners from Cornwall, UK.

Woody debris: branches, trunks, roots, and whole trees—including units a metre in diameter or more—that fall into and can be carried along in rivers or trapped in log jams. Common now along parts of Irvine Creek, rare in the Grand, but once probably a major factor everywhere before the postglacial forest cover was removed.

References

RECOMMENDED READING

Brintnell, Candace. "Architecture and Stratigraphy of the Lower Silurian Guelph Formation, Lockport Group, Southern Ontario and Michigan." MSc thesis, University of Western Ontario, 2012. University of Western Ontario Electronic Thesis and Dissertation Repository, https://ir.lib.uwo.ca/etd/632.

Chapman, L.J. and D.F. Putnam. *The Physiography of Southern Ontario*. Ontario Geological Survey Special Volume 2. 3rd ed. Toronto: Ontario Ministry of Natural Resources, 1984.

Eyles, Nick. Ontario Rocks: Three Billion Years of Environmental Change. Markham, ON: Fitzhenry & Whiteside, 2002.

Greenhouse, J.P., and P.F. Karrow. "Geological and Geophysical Studies of Buried Valleys and their Fills near Elora and Rockwood, Ontario." *Canadian Journal of Earth Sciences* 31, no. 12 (December 1994): 1838-1848. https://doi.org/10.1139/e94-163.

Karrow, P.F. *Pleistocene Geology of the Guelph Area, Southern Ontario*, Geological Report 61. Toronto: Ontario Department of Mines, 1968. http://www.geologyontario.mndm.gov.on.ca/mndmaccess/mndm_dir.asp?type=pub&id=R061.

Phillips, R.T.J., and J.R. Desloges. "Glacial Legacy Effects on River Landforms of the Southern Laurentian Great Lakes." *Journal of Great Lakes Research* 41, no. 4 (December 2015): 951-964. https://doi.org/10.1016/j.jglr.2015.09.005.

Tinkler, Keith J., and Ellen E. Wohl, eds. *Rivers over Rock: Fluvial Processes in Bedrock Channels*, Geophysical Monograph 107. Washington, DC: American Geophysical Union, 1998.

OTHER WORKS CITED

Allan, Roberta, compiler. *History of Elora*. Elora, ON: Elora Women's Institute, 1982.

Anderson, Elizabeth P., Catherine M. Pringle, and Mary C. Freeman. "Quantifying the Extent of River Fragmentation by Hydropower Dams in the Sarapiquí River Basin, Costa Rica." *Aquatic Conservation: Marine and Freshwater Ecosystems* 18, no. 4 (June 2008): 408–417. https://doi.org/10.1002/aqc.882.

Boyd, Dwight, Sam Bellamy, Dave Schulz, and Gordon Nelson. "The Water Regime or Hydrology of the Grand River Watershed." In Nelson, *Towards a Grand Sense of Place,* 47–58.

Boyle, David. *On the Local Geology of Elora,* Selected Papers from Proceedings of the Elora Natural History Society, 1874–5. Elora, ON: J. Townsend, 1875. https://archive.org/details/selectedpapers00enhsuoft/mode/2up.

Briggs, Louis I., and Darinka Briggs, *Niagara-Salina Relationships in the Michigan Basin*. Lansing, MI: Michigan Basin Geological Society, 1974.

Brilha, José, and Emmanuel Reynard. "Geoheritage and Geoconservation: The Challenges." In *Geoheritage:*

Assessment, Protection, and Management, ed. Emmanuel Reynard and José Brilha, 433–38. Amsterdam: Elsevier, 2018. https://doi.org/10.1016/C2015-0-04543-9.

Brunton, F.R. "Project Unit 08-004. Update of Revisions to the Early Silurian Stratigraphy of the Niagara Escarpment: Integration of Sequence Stratigraphy, Sedimentology and Hydrogeology to Delineate Hydrogeologic Units." In *Summary of Field Work and Other Activities 2009*, Ontario Geological Survey, Open File Report 6240, 25-1–25-20. Sudbury, ON: Ministry of Northern Development and Mines, 2009.

Brunton, F.R., and J.E.P. Dodge. *Karst of Southern Ontario, Including Manitoulin Island*. Ontario Geological Survey, Groundwater Resources Study 5. Sudbury, ON: Ontario Ministry of Northern Development and Mines, 2008. http://www.geologyontario.mndm.gov.on.ca/mndmaccess/mndm_dir.asp?type=pub&id=GRS005 (last modified January 20, 2017).

Brunton, F.R., L. Smith, O.A. Dixon, P. Copper, H. Nestor, and S. Kershaw. "Silurian Reef Episodes, Changing Seascapes and Paleobiogeography." In *Silurian Cycles: Linkages of Dynamic Stratigraphy with Atmospheric, Oceanic, and Tectonic Changes*. Proceedings of the Second International Symposium on the Silurian System, New York State Museum Bulletin 491, 259–76. Albany, NY: State University of New York, State Education Department, 1998.

Burr, Devon M., Paul A. Carling, and Victor R. Baker, eds. *Megaflooding on Earth and Mars*. Cambridge: Cambridge University Press, 2009.

Carcavilla, Luis, Juan José Durán, Ángel García-Cortés, and Jerónimo López-Martínez. "Geological Heritage and Geoconservation in Spain: Past, Present, and Future." *Geoheritage* 1 (October 2009): 75–91. https://doi.org/10.1007/s12371-009-0006-9.

Carling, Paul A., Mauricio Perillo, Jim Best, and Marcelo H. Garcia. "The Bubble Bursts for Cavitation in Natural Rivers: Laboratory Experiments Reveal Minor Role in Bedrock Erosion." *Earth Surface Processes and Landforms* 42, no. 9 (July 2017): 1308–16. https://doi.org/10.1002/esp.4101.

Church, Michael. "Bed Material Transport and the Morphology of Alluvial River Channels." *Annual Review of Earth and Planetary Sciences* 34 (May 2006): 325–54. https://doi.org/10.1146/annurev.earth.33.092203.122721.

Cole, Jason, Mario Coniglio, and Simon Gautrey. "The Role of Buried Bedrock Valleys on the Development of Karstic Aquifers in Flat-Lying Carbonate Bedrock: Insights from Guelph, Ontario, Canada." *Hydrogeology Journal* 17, no. 6 (September 2009): 1411–25. https://doi.org/10.1007/s10040-009-0441-3.

Coniglio, Mario, Qing Zheng, and Terry R. Carter. "Dolomitization and Recrystallization of Middle Silurian Reefs and Platformal Carbonates of the Guelph Formation, Michigan Basin, Southwestern Ontario." *Bulletin of Canadian Petroleum Geology* 51, no. 2 (June 2003): 177–99. https://doi.org/10.2113/51.2.177.

Connon, John Robert. *The Early History of Elora and Vicinity.* 1930. Reprinted with an introduction by Gerald Noonan. Waterloo, ON: Wilfrid Laurier University Press, 1975.

Crutzen, Paul J. "The 'Anthropocene.'" In *Earth System Science in the Anthropocene: Emerging Issues and Problems*, ed. Eckart Ehlers and Thomas Krafft, 13–18. Berlin: Springer, 2006.

Curran, Janet H., and Ellen E. Wohl. "Large Woody Debris and Flow Resistance in Step-Pool Channels, Cascade Range, Washington." *Geomorphology* 51,

nos. 1–3 (March 2003): 141–57. https://doi.org/10.1016/
S0169-555X(02)00333-1.

Dynesius, Mats, and Christer Nilsson. "Fragmentation and
Flow Regulation of River Systems in the Northern Third
of the World." *Science* 266, no. 5186 (November 1994):
753–62. https://doi.org/10.1126/science.266.5186.753.

Elora Victoria Crescent Neighbourhood Heritage Conservation District Study Subcommittee. *Elora's Victoria Crescent Neighbourhood Heritage Conservation District Study.*
Elora, ON: Elora Victoria Crescent Neighbourhood Heritage Conservation District Study Subcommittee, 2010.

Eyles, Nicholas, Emmanuelle Arnaud, Adrian E. Schei-
degger, and Carolyn H. Eyles. "Bedrock Jointing and
Geomorphology in Southwestern Ontario, Canada:
An Example of Tectonic Predesign." *Geomorphology*
19, nos. 1–2 (May 1997): 17–34. https://doi.org/10.1016/
S0169-555X(96)00050-5.

Ford, Derek C., and Paul Williams. *Karst Geomorphology
and Hydrology.* Rev. ed. Hoboken, NJ: Wiley, 2007.

Gao, Cunhai. "Buried Bedrock Valleys and Glacial and Subglacial Meltwater Erosion in Southern Ontario, Canada."
Canadian Journal of Earth Sciences 48, no. 5 (May 2011):
801–18. https://doi.org/10.1139/e10-104.

Geological Society of America. *GSA Position Statement:
Geoheritage.* Boulder, CO: Geological Society of America,
2017. https://www.geosociety.org/gsa/positions/position
20.aspx.

Grand River Conservation Authority. "Elora Gorge."
Accessed May 15, 2023. https://www.grandriver.ca/en/
outdoor-recreation/Elora-Gorge.aspx.

Hackett, Sophie, Andrea Kunard, and Urs Stahel, eds.
Anthropocene: Burtynsky, Baichwal, de Pencier. Toronto:

Art Gallery of Ontario; Fredericton, NB: Goose Lane Editions, 2018. Exhibition catalogue.

Harp, Edwin L., and Randall W. Jibson. "Anomalous Concentrations of Seismically Triggered Rock Falls in Pacoima Canyon: Are They Caused by Highly Susceptible Slopes or Local Amplification of Seismic Shaking?" *Bulletin of the Seismological Society of America* 92, no. 8 (December 2002): 3180–89. https://doi.org/10.1785/0120010171.

Henry, Michael, and Peter Quinby. *Ontario's Old-Growth Forests*. 2nd ed. Markham, ON: Fitzhenry & Whiteside, 2021.

Hewitt, Kenneth. *Elora Gorge: A Visitor's Guide*. Toronto, ON: Stoddart, 1995.

———. "Human Society as a Geological Agent." In *Quaternary Geology of Canada and Greenland*, Geology of Canada No.1, ed. R.J. Fulton, 624–34. Ottawa, ON: Geological Survey of Canada, 1989. https://doi.org/10.4095/127905.

Janzen, Bob. *Geology of the Grand River Watershed: An Overview of Bedrock and Quaternary Geological Interpretations in the Grand River Watershed*. Cambridge, ON: Grand River Conservation Authority, 2018. https://www.grandriver.ca/en/our-watershed/resources/Documents/Groundwater/Watershed-Geology_March272019.pdf.

Jennings, J.N. *Karst Geomorphology*. 2nd ed. New York: Blackwell, 1985.

Karrow, P.F. "Quaternary Geology of the Great Lakes Subregion." In *Quaternary Geology of Canada and Greenland*, Geology of Canada No. 1, ed. R.J. Fulton, 326–50. Ottawa, ON: Geological Survey of Canada, 1989.

Kehew, Alan E., Mark L. Lord, Andrew L. Kozlowski, and Timothy G. Fisher. "Proglacial Megaflooding along the

Margins of the Laurentide Ice Sheet." In Burr et al., *Mega-flooding on Earth and Mars*, 104–27.

Killan, Gerald. *David Boyle: From Artisan to Archaeologist*. Toronto, ON: University of Toronto Press, 1983.

Logan, William E. *Geology of Canada: Geological Survey of Canada—Report of Progress from its Commencement to 1863*. Montreal: Dawson Bros., 1863.

Martin, Virgil. "Grand River Vegetation and Wildlife." In Nelson, *Towards a Grand Sense of Place,* 59–83

Martini, I.P. "Gravelly Flood Deposits of Irvine Creek, Ontario, Canada." *Sedimentology* 24, no. 5 (October 1977), 603–22. https://doi.org/10.1111/j.1365-3091.1977 .tb00260.x.

McNeill, J. R., and Peter Engelke. *The Great Acceleration: An Environmental History of the Anthropocene since 1945*. Cambridge, MA: Harvard University Press, 2014.

Morgan, Alan V., and Paul F. Karrow. "Bedrock Geology of the Grand River Watershed." In Nelson, *Towards a Grand Sense of Place,*11–20.

Munro-Stasiuk, Mandy J., John Shaw, Darren B. Sjogren, Tracy A. Brennand, Timothy G. Fisher, David R. Sharpe, Philip S.G. Kor, Claire L. Beaney, and Bruce B. Rains. "The Morphology and Sedimentology of Landforms Created by Subglacial Megafloods." In Burr et al., *Megaflooding on Earth and Mars*, 78–103.

Murray, Alexander. "Report of Alex. Murray, Esq., Assistant Provincial Geologist, addressed to W. E. Logan, Esq., Provincial Geologist." In *Geological Survey of Canada: Report of Progress for the Year 1850–51*, ed. William E. Logan (Quebec: John Lovell, 1852), 13–33. https://doi .org/10.4095/123558.

Nelson, Gordon, ed. *Towards a Grand Sense of Place: Writings on Changing Environments, Land-Uses, Landscapes, Lifestyles and Planning of a Canadian Heritage River.* Waterloo Department of Geography Publications Series. Waterloo, ON: Heritage Resources Centre, University of Waterloo, 2004.

Nilsson, Christer, Catherine A. Reidy, Mats Dynesius, and Carmen Revenga. "Fragmentation and Flow Regulation of the World's Large River Systems." *Science* 308, no. 5720 (April 2005): 405–08. https://doi.org/10.1126/science.1107887.

Ontario Geological Survey, *Bedrock Geology of Ontario, Southern Sheet.* 1991. 1:1,000,000 scale, Map 2544. Ontario Ministry of Northern Development, Mines, Natural Resources and Forestry, Surficial Geology Series, http://www.geologyontario.mndm.gov.on.ca/mines/ogs/indexes/maps_e.html.

Parks, W.A. *The Stromatoporoids of the Guelph Formation in Ontario.* University of Toronto Studies, Geological Series, No. 4. Toronto: University of Toronto, 1907. https://archive.org/details/cihm_80863.

Sanderson, Marie. *The Grand Climate: Weather and Water in the Grand River Basin.* Cambridge, ON: Grand River Foundation, 1998.

Sanderson, Marie, and Brian Mills. "Climate Variability in the Grand River Watershed." In Nelson, *Towards a Grand Sense of Place,* 35–46.

Sanford, B.V. "Silurian of Southwestern Ontario." In Ontario Petroleum Institute, Proceedings of the 8th Annual Conference, Technical Paper No. 5, 1–44. London, ON: Ontario Petroleum Institute, 1969.

Selby, M.J. *Hillslope Materials and Processes*. 2nd ed. Oxford: Oxford University Press, 1993.

Seyfert, Carl K., ed. *Encyclopedia of Structural Geology and Plate Tectonics*. New York: Van Nostrand Reinhold, 1987.

Shtober-Zisu, Nurit, Hani Amasha, and Amos Frumkin. "Inland Notches: Implications for Subaerial Formation of Karstic Landforms—An Example from the Carbonate Slopes of Mt. Carmel, Israel." *Geomorphology* 229 (January 2015): 85–99. https://doi.org/10.1016/j.geomorph.2014.09.004.

Steelman, Colby M., Emmanuelle Arnaud, Peeter Pehme, and Beth L. Parker. "Geophysical, Geological, and Hydrogeological Characterization of a Tributary Buried Bedrock Valley in Southern Ontario." *Canadian Journal of Earth Sciences* 55, no. 7 (July 2018): 641–58. https://doi.org/10.1139/cjes-2016-0120.

Steffen, Will, Paul J. Crutzen, and John R. McNeill. "The Anthropocene: Are Humans Now Overwhelming the Great Forces of Nature?" *Ambio* 36, no. 8 (December 2007): 614–21. https://doi.org/10.1579/0044-7447(2007)36[614:TAAHNO]2.0.CO;2.

Templin, Hugh. *Fergus: The Story of a Little Town*. Fergus, ON: Fergus News-Record, 1933.

Thorning, Stephen. "Fish, clean rivers a concern in Wellington 165 years ago." *Wellington Advertiser*, April 19, 1999; repr. May 20, 2021, https://www.wellingtonadvertiser.com/fish-clean-rivers-a-concern-in-wellington-165-years-ago/.

———. "Conservation initiatives of 1913 failed to get action." *Wellington Advertiser*, September 22, 2006; repr. July 15, 2020, https://www.wellingtonadvertiser.com/conservation-initiatives-of-1913-failed-to-get-action/.

Tinkler, Keith J. "Fluvially Sculpted Rock Bedforms in Twenty Mile Creek, Niagara Peninsula, Ontario." *Canadian Journal of Earth Sciences* 30, no. 5 (May 1993): 945–53. https://doi.org/10.1139/e93-079.

Tucker, Maurice E., and V. Paul Wright. *Carbonate Sedimentology.* Malden, MA: Blackwell Scientific, 1990.

Warrick, Gary. "Early Aboriginal Occupation of the Grand River Watershed." In Nelson, *Towards a Grand Sense of Place*, 83–95.

Watts, Vanessa. "Indigenous Place-Thought and Agency amongst Humans and Non-Humans (First Woman and Sky Woman Go on a European World Tour!)." *Decolonization: Indigeneity, Education and Society* 2, no. 1 (May 2013): 20–34. https://jps.library.utoronto.ca/index.php/des/article/view/19145.

Weston, Thomas Chesmer. *Reminiscences Among the Rocks in Connection with the Geological Survey of Canada.* Toronto, ON: Warwick Bros & Rutter, 1899.

Williams, M.Y. *The Silurian Geology and Faunas of Ontario Peninsula, and Manitoulin and Adjacent Islands.* Memoir 111, Geological Series No. 91. Ottawa, ON: Geological Survey of Canada, 1919. https://archive.org/details/cihm_82237.

Wohl, Ellen E. "Bedrock Channel Morphology in Relation to Erosional Processes." In Tinkler and Wohl, *Rivers over Rock*, 133–51.

———. *Mountain Rivers Revisited*, Water Resources Monograph 19. Washington, DC: American Geophysical Union, 2010.

———. "Spatial Heterogeneity as a Component of River Geomorphic Complexity." *Progress in Physical Geography* 40, no. 4 (August 2016): 598–615.

Wohl, Ellen E., Brian P. Bledsoe, Robert B. Jacobson, N. Leroy Poff, Sara L. Rathburn, David M. Walters, and Andrew C. Wilcox. "The Natural Sediment Regime in Rivers: Broadening the Foundation for Ecosystem Management." *BioScience* 65, no. 4 (April 2015): 358–71. https://doi.org/10.1093/biosci/biv002.

Zeller, Suzanne. *Inventing Canada: Early Victorian Science and the Idea of a Transcontinental Nation.* Toronto, ON: University of Toronto Press, 1987.

Acknowledgements

I am indebted to certain neighbours, friends, and colleagues living near the gorge, and for the schooling they provided in their recollections from countless walks; in particular, the late Beverley Cairns, Graeme Cairns, Donnie Eaton, Elizabeth Fasken, Peter Merry, and the late David Wilcox. There have been instructive experiences on many field trips from WLU, and half a dozen field excursions with colleagues and visiting scientists including exchanges with Monique Fort, Edward Relph, and Houston Saunderson.

Early on, I learned much that was new to me from conversations with and newspaper articles by the late Stephen Thorning, a specialist on the history of this region. I was also assisted by H. Julia Roberts, who conducted archival research into records of early geologists in the region, under the supervision of historian Dr. Suzanne Zeller. A number of works on the area are valuable for local histories, rich in personalities, anecdotes, and local events (Weston 1899; Connon [1930] 1975; Allan 1982). Like most earth scientists working in Southern Ontario, I am beholden to the painstaking reconstructions of the landscapes by L.J. Chapman and D.F. Putnam. I am indebted to a number of other researchers who study the bedrock, including F.R. Brunton, Candace Brintnell, and especially Paul Karrow's work on Quaternary developments in the region, as well as others listed in the References. This said, it must also be noted that past work generally has concentrated on other areas or local subjects besides Elora Gorge itself. The nature and story of the gorge landscapes have received very limited attention.

I am especially indebted to Stepan Wood for editing the manuscript. Most of the maps and diagrams were prepared by the cartographer Pam Schaus. Thanks to Sara Wood for drawings of fossils found in Elora bedrock, to Michael Bach-meier, Tara Hewitt, and Nina Hewitt for accompanying me on excursions to the gorge to take new photographs for this edition, and to Nina Hewitt for help selecting photographs for inclusion. Photographs are my own unless otherwise indi-cated. Thanks are also due to the Wellington County Museum and Archives for help locating archival photographs. Finally, I am grateful to two anonymous reviewers for valuable com-ments on the original manuscript, and to everyone at the Wil-frid Laurier University Press for helping to bring this project to fruition.

Index

accordant junction, 126, 201
   See also Grand-Irvine junction
"Ages of the Gorge" (box), 166–67
aggregate resources, 176
Algonquin Arch, 64f, 81, 83–84
   See also "sleeping giants"
Algonquin Lake, 98
alluvium, 8, 201
anchor ice, 123, 201
Anthozoa, 71
Anthropocene, 103, 167, 177–183
aragonite, 74, 191n13, 201, 202

bafflers, 74
Bahamas, 68, 97
bare karst, 154, 202
barrier reefs
   of Australia, 68
   of Bahamas, 68
   defined, 202
   formation and features of, 69–70, 78–79, 81–83, 152
   Guelph Formation as, 40–41, 68, 70–71, 166
   strength of, 142

   See also bedrock; coral reefs; fossils; reef-building organisms
BBVs. *See* buried bedrock valleys (BBVs)
beavers, 101, 121, 157, 161, 167
bedrock
   control of landforms, 8–9, 41, 66, 95, 105, 107, 163
   erosion of rockwalls, 133–34
   exposed in Elora area, 7–8, 81–82
   fluvial erosion of, 105, 107–14, 116–19
   formation and features of, 22, 63–68, 83–85, 107, 166
   fractures and falls, 141, 144–48
   photographs of, 2, 20, 50, 70, 104, 109, 111, 135
   of Southern/Southwestern Ontario, 22, 64f, 83, 84f
   See also barrier reefs; buried bedrock valleys (BBVs); dolostone; Elora Gorge; Guelph Formation; Silurian Period